FROM PHOTONS TO ATOMS

The Electromagnetic Nature of Matter

FROM PHOTONS TO ATOMS

The Electromagnetic Nature of Matter

Daniele Funaro
Università di Modena e Reggio Emilia, Italy

World Scientific

NEW JERSEY · LONDON · SINGAPORE · BEIJING · SHANGHAI · HONG KONG · TAIPEI · CHENNAI · TOKYO

Published by

World Scientific Publishing Co. Pte. Ltd.
5 Toh Tuck Link, Singapore 596224
USA office: 27 Warren Street, Suite 401-402, Hackensack, NJ 07601
UK office: 57 Shelton Street, Covent Garden, London WC2H 9HE

British Library Cataloguing-in-Publication Data
A catalogue record for this book is available from the British Library.

FROM PHOTONS TO ATOMS
The Electromagnetic Nature of Matter

ISBN 978-981-120-423-4

For any available supplementary material, please visit
https://www.worldscientific.com/worldscibooks/10.1142/11383#t=suppl

Typeset by Stallion Press
Email: enquiries@stallionpress.com

Preface

The aim of this exposition is to informally present my viewpoint concerning "matter" and the way it is structured. The analysis is intended to reveal and better understand the secrets of the basic components of our universe. I will try to minimize the use of jargon, with the hope of making this material accessible to readers with well developed scientific backgrounds, but without specific professional skills. Most of the sentences reported here are, however, consequences of a solid mathematical model, that combines together in an elegant way and with a relatively simple appearance, well-established results that are recognized world-wide to form the backbone of classical physics. Regarding a certain number of questions, I am unable to come up with a mathematical proof. In this case, my conclusions will be supported only by the experience gained in these last years. I am confident that precise answers, not too far from the guessed ones, will be devised sooner or later. Consequently this collection of facts may be improved and updated as additional material and confirmation become available.

Let me briefly mention some justifications that are at the foundation of my philosophy. Some have been expressively manifested in my previous book ([Funaro(2008)]) and the successive papers, some others are only vaguely anticipated therein. These arguments slowly took a global and organic form in my mind, although, since the very beginning, there was an underlying well-defined pattern, the guidelines of which will be discussed in the pages to follow.

The material has been carefully collected and developed in order to adhere to the existing evidence, documented by a multitude of publications and experiments. In this regard let me point out that the possibility to access public information through the instruments of modern informatics, and the ability of creating suitable connections, are precious tools when sagely used, having helped my research a lot. The determination of not being influenced by known and established theories has also been an important factor. In fact, the search for a common denominator, based on an Occam's razor procedure that led to an overall solution, had to prevail over immediate local explanations, acceptable to the first approximation, but irreparably false when extended to cover contiguous subjects. Another stimulus came from the necessity of describing things in a straightforward way, as the perception of our senses (helped, if needed, by the use of sophisticated instruments) would suggest. To better clarify this concept, I did not want to invoke, for example, the existence of unnatural environments, such as additional space dimensions, parallel universes or a general entanglement of all bodies through a non well-clarified mechanism. One of the assumptions was that the constituting rules had to be in line with our way of decoding external stimulations. Hence, matter had not to be more complicated than the minimum allowed to originate the phenomena that fill up our everyday life.

Of course, these concepts were present and clear up to a century ago, until some controversial experiments revealing the quantum structure of matter brought disorder in the scientific community. Ingenious explanations were given, but this slowly led to a radical change in the way of interpreting nature, so that nowadays the general approach seems not to fit common sense. Thus, why should I be able, despite the efforts of distinguished scientists, to restore the old-fashioned vision and come out with reliable answers? Well, first of all, many decades have passed and numerous other experiments have disclosed new results, not even expected before. These new insights certainly provide some of the missing pieces of the puzzle. The important thing, as I mentioned above, is to allow loose ends and not to rush to conclusions. In addition to this sort of road map, I could count on two decisive facts that definitely supported the development of my thoughts.

The first important step is the review of electromagnetism actuated in [Funaro(2008)]. Assimilating those results is quite crucial, since, hidden in crude formulas, something innovative lies. My intent here is to reveal in a painless way the secret path leading to the comprehension of photons, because, successively, it will be the cantilever to access the atom's world. The skeptical reader may think that there is no need to propose additional theory for electromagnetic waves, because the domain of applications does not seem to require it. However, considering that the known approaches are not even capable of simulating basic entities, like spherical or solitary waves, the new model is intended to be complementary, rather than alternative to the existing one. I will do my best to set forth these considerations with rigorous arguments.

In truth, my job has been just a scrupulous re-reading of existing material, so that my model merges Maxwell's and shares with it all the good features. This revolution may still not be considered decisive for everyday large-scale applications, but it is certainly unavoidable when one tries to describe what really happens at the atomic and molecular level. A convincing understanding of my theoretical arguments can only be appreciated after clearing the path from preconceived schemes, vaguely formalized in the early discoveries and inherited as incontrovertible truths. By saying this, I do not lack respect for the pioneering work of J. C. Maxwell, who, inaugurating a new era of scientific knowledge, was quite conscious (more than many of his successors) of the physical meaning of the results he was achieving[1].

The second point in favor of my theory is the empirical evidence of a large amount of electromagnetic energy coexisting with matter. In quantum mechanics this is called zero-point radiation and it is also present in pure vacuum. Officially, it comes from the uncertainty in the description of physical events in the very small scale. This may sound quite normal if one thinks according to a probabilistic description of nature, where some fuzziness is attributed to matter. It is not a convincing explanation when one is looking for a classical type answer, based on a set of deterministic differential equations, where cause and effect follow a rigorous stream. As I will describe in the coming chapters, a deeper knowledge of the properties of this radiation is unavoidable to arrive at a full description of the structure of matter, because this energy is not just an innocent by-product, but the primary ingredient of our universe.

As I have already said, the exposition given here is mainly qualitative. Therefore, some known facts will be reported in a simplistic way, though they are based on the work of many excellent researchers. The dissertation is organized in three chapters, plus a further one, where technical math-

---

[1] In [Maxwell Commemorative Booklet(1999)], K. Moffatt writes: "The distillation of the great mass of experimental knowledge accumulated during the earlier decades of the Nineteenth Century into what we now know as Maxwell's equations provides evidence of real genius; the beauty, and ultimate simplicity, of these equations is still a matter of wonder for students of applied mathematics in our universities, and recognition of the significance of Maxwell's equations must count as one of the high points of any degree course in mathematics and physics".

ematical issues are listed. The first chapter, dealing with the modeling equations, is addressed to readers with a slightly more specialized preparation. I expose my motivations by skipping technicalities, continuing the investigation started in my previous book. This preliminary reading is crucial to appreciate the spirit of the successive chapters; its full assimilation is however not necessary in order to proceed. The second chapter deals with elementary particles and subnuclear structures. The outcome is, as much as I could, in line with the experiments and the guidelines of my model. Reaching more robust conclusions is a matter of intensive investigation, mainly in the field of numerical simulations. Covering with due care such an extended and controversial area of knowledge is impossible. Therefore I opted for a general overview that certainly needs debugging, but which points to an escape route from the dark corridor of modern nuclear physics. The third chapter is about the study of basic atoms and chemical structures. The main battleground is to provide explanations on the significance and the origin of the quantum properties of matter. Again, due to scarcity of time, my role is limited to the collection of a series of facts, aimed at revealing their common underlying nature.

Let me conclude this introduction by remarking that the most relevant achievement of my analysis is its philosophical implication. This paper is not only intended to be just an explanation of mechanisms, but rather an essay on a possible way to look through different eyes at the universe we live in.

D. Funaro

BRIEF SUMMARY

• Lights phenomena, propagating at finite speed, have a tendency to create zones, within the electromagnetic radiation, where the divergence of the electric field is different from zero, even in absence of genuine material charges.

• A review of the model equations describing electromagnetic phenomena in vacuum is then necessary, in order to take into consideration the occurrence of nonvanishing charge densities.

• A clever mix of the classical equations ruling the dynamics of inviscid fluids and Maxwell's equations, provides the right model, satisfying all the prerogatives normally requested by theoretical physics. In a preliminary analysis, the model does not account for moving and interacting "solid" charged particles, but only deals with electromagnetic fields considered as primitive entities.

• The velocity field which is part of the model is not related to the mechanical movement of massive particles, but to the way the electromagnetic information develops. A scalar potential, that may acquire the role of "pressure", is also present in the equations.

• The new set allows for an impressively large space of solutions, containing waves traveling according to the rules of geometrical optics (not present in Maxwell's model). In particular, among the solutions that can be explicitly written, there are electromagnetic waves displaying compact support and traveling straightly and undisturbed at the speed of light. This opens the path to the explanation in classical terms of the wave-particle dualism.

• Particle-waves, of the kind described above, modify the structure of space-time. Explicit solutions of Einstein's equations are available in the most elementary situations. These are obtained by plugging the electromagnetic stress tensor into the right-hand side.

• As it may happen in fluid dynamics, electromagnetic waves can form stable structures. Vortex rings are the most peculiar geometries. In the context of general relativity, stable configurations are the result of a space-time modification due to the periodic behavior of the waves, that, mixing cause and effect, follow the same closed geodesics generated by their evolution.

• Stable electromagnetic rings displaying all the features of electrons (size, charge. mass, magnetic properties) are workable solutions of the revised model. An electron is not a point-wise entity anymore. Spin corresponds to an actual rotation, realized in the fashion of vortex rings, that do not revolve around an axis.

• With the help of some inventiveness, other particles (stable or not stable), such as the muon, the proton, or the neutron, can be built according to similar principles. In this fashion, quarks and gluons are not particles, but electromagnetic manifestations self-trapped in tiny regions of space. Nevertheless, if suitably reinterpreted, the axioms of the principle of indeterminacy still hold.

• Through an appropriate change of sign in the model equations, anti-particles can be rediscovered. This construction reinforces the conjecture of CPT invariance. One can then modify Parity, passing from a universe into its anti-universe either by switching the Charge or by reversing the arrow of Time. In basic situations, waves and anti-waves are specular entities. Regarding particles and anti-particles the distinction is more subtle.

• Subatomic particles (or anti-particles) end up to be isolated geometrical objects. They may enter in communication only when immersed in an electromagnetic background filling up the whole universe. Again, the model equations are able to describe such a dynamical flow, with no need to introduce concepts such as aether. As in general relativity, the universe is a global entity, and particles live in special self-confined domains (having the same electromagnetic nature of the surroundings), without the necessity of being concentrated singularities The gravitational world (interpreted as space-time deformation) and the electromagnetic one are now indissolubly fused together. Space-time curvature is strictly related to the activation of the pressure term in the equations.

• Nuclei of atoms, obtained from aggregation of elementary particles, have strong influence on the surrounding electromagnetic sea. This comes as a consequence of their electrical and magnetic properties, that are transferred and shared with other nuclei or spare electrons. Spin is also an important factor in this process.

• Atoms and molecules are constituted through the mediation of the electromagnetic vehicle, by exerting reciprocal forces that go far beyond the classical Coulomb's law. There is no reason at this point to assume that electrons revolve around nuclei, while photon emission is clearly explained through the breakage of electromagnetic links during atom excitation. Molecules are in this way the blending of effective matter (particles) and the energy connecting them. Such a union has a conformation that does not necessarily reflect those pertaining to the single parts of the ensemble.

• Under this construction, the structure of our Universe turns out to be of pure electromagnetic nature. Traveling photons, standard elementary particles and nuclei can be experimentally detected; so they are easily recognized. Nevertheless, a large amount of energy is distributed in the whirling streams of an electromagnetic flow, presenting an extended range of frequencies that decay with the distance from the sources. High frequencies are located within the nuclei (gamma rays) or immediately around them (X-rays), going down to visible photons as dimensions increase, up to the very low frequencies of the astronomic environment.

• Such an organized universe is a non-smooth dynamical continuum (with no concentrated singularities, however) that reveals its features through a myriad of quantized substructures. The important fact is that quantization is not a peculiarity of the model equations, but of the solution's behavior.

• This "holistic" view allows for a comprehensible and coherent explanation of many quantum phenomena and gives access to a more sound analysis of the relationships between biology and electromagnetism (biophotonics), including, more in general, the liaison between humans and the surrounding world.

Contents

Chapter 1

The World of Photons

Gli è tutto sbagliato, tutto da rifare
Gino Bartali, cyclist

1.1 Why improve the theory of electromagnetism?

The fundamental brick of my construction process is the *photon*. Somebody may find more appropriate the terms *solitary wave* or *soliton*. Although physicists are quite sensitive about the correct nomenclature, from the developments of this chapter it will come out that there is actually no distinction between the various concepts. I will remark however when some substantial differences may emerge.

Photons are ubiquitous. Nevertheless, experts can only count on vague informal definitions about photons, whereas they have at their disposal very accurate descriptions regarding their behavior. According to quantum theories, photons are the *carriers* of light and all electromagnetic phenomena in general. In other words, a photon is supposed to be an elementary particle. It is the "quantum" of the electromagnetic field, the basic unit of all forms of electromagnetic radiation. The photon has no rest mass and exhibits properties of both waves and particles. The very last statement has been, and still is (despite an unconscious attitude to hide the problem), one of the crucial questions in modern physics. I will devote here my efforts to come out with convincing explanations aimed at giving meaning to the ambiguity of the expression "wave-particle". The endeavor is of paramount importance, because it will bring me to decrypt well-known facts from quite a different perspective, providing a new interpretation of elementary concepts such as matter and energy, and studying the way they transform into each other.

The above "definitions" of photon are clearly unsatisfactory[2,3]. Describing such solitary *wave-packets* in a classical context, that is by means of a set of partial differential equations, where the unknowns give an exact account of the position and the dynamics, has been the goal of many researchers for more than a century. The reason for this difficulty can be ascribed to the standard Maxwell's model, initially supposed to provide full explanation of electromagnetic phenomena, but resulting in failure as the first quantum problems emerged, turning out to be totally unsuited to the treatment of photons.

For a technical review of Maxwell's equations in vacuum and their main properties, the reader is referred to appendix A (and, of course, to all specialized publications, starting from basic textbooks). For an account of the properties of photons, within the quantum and the semiclassical approach, the reader is addressed to [Keller(2005)] and [Kidd, Ardini and Anton(1989)], as well as to the references cited there. The large amount of material available should not lead us to think that the topic is trouble free. Although seemingly complete, work on photons suffers from problems at a foundational level, usually well hidden under a thick carpet, notwithstanding that practical consequences of the existing models are largely correct up to a certain degree of accuracy.

I claim to have achieved the result of fully incorporating photons within the framework of differential models (see [Funaro(2008)] and the earlier paper [Funaro(2005)]), but it has been necessary to sacrifice some aspects, taken for granted, of Maxwell's original model, and this is of course a source of controversy. I seriously believe that, if the revolution process had started a long time ago, physics would have evolved along other paths. Unfortunately, many conceptual approaches have survived for generations and are so established in the scientific community that it is hard to propose alternatives. In addition to this, one has to account for inaccuracies and mistakes, sometimes handed down through generations of students as urban myths,

---

[2]From [Roychoudhuri el al.(2008)], p.3 (A. Zajonc): "What are light quanta? Of course today every rascal thinks he knows the answer, but he is deluding himself. We are today in the same state of 'learned ignorance' with respect to the light as was Einstein".

[3]From [Beil(1997)]: "It is common now to argue that photons are not real particles at all and should not be subject to visualizable models. This is the attitude of most interpretations of quantum electrodynamics. Photons are supposed to be only the quantized excitations of the normal modes of the system".

that have hindered the search for a rigorous framework[4]. A responsibility may be in part attributed to mathematicians for having underestimated the importance of the problem[5]. Many of the things I am going to say may sound unreliable and unexpected to an engineer and, at the same time, trivial and unessential to a theoretical mathematician. My role here is in trying to match both viewpoints. Since this kind of "in between" discipline is not officially coded and recognized, I expect to be criticized on both fronts.

Let me warn the reader that if he wants to appreciate the results of my work it is necessary to cast any passage within a general framework, i.e., the validation of the theory is a process to be taken as a whole and judged at the end of the exposition. Any premature attempt to jump into conclusions on the basis of already experimented schemes will be biased towards the search of a local answer, while, as I specified in the introduction, we are trying to solve an entire puzzle. I know that there are readers ready to shoot me at the first mistake or incorrect statement. I ask them to be patient and let me survive at least up to the end of the story.

The first important achievement of the modified equations introduced in [Funaro(2008)] is that they make no actual difference between a photon and any other electromagnetic phenomena (as, for instance, waves emanating from an antenna), since they all result from manifestations of the same unifying theory. With rough approximation the reader may find it convenient to consider a 2D wave like that produced by a pebble thrown vertically into water (see figure 1.3). Later, it will be compulsory to investigate more properly the internal nature of these wakes. As far as I am concerned, the evolution of an electromagnetic wave is exclusively ruled by geometrical laws. This is going to be true at least for those configurations that will be called *free-waves* (see section 1.3). More precisely, the rules are those of of *geometrical optics*, though it is necessary to extend the concept of space as done in *general relativity* (I will better specify this point

---

[4]For example, in [Oppenheimer(1989)], p.113, we read: "A typical electromagnetic wave may have the electric force changing with time periodically, [...] the magnetic force is doing the same thing at right-angles to the electric force and out of phase with it, so that when the electric force is zero the magnetic force is a maximum ...". The sentence, although made by an authoritative fellow, is irreparably wrong.

[5]As stated in [Dyson(1972)]: "The first clear sign of a breakdown in communication between physics and mathematics was the extraordinary lack of interest among mathematicians in James Clerk Maxwell's discovery of the laws of electromagnetism. [...] I shall try to convince you by examining actual cases that the progress of both mathematics and physics has in the past been seriously retarded by our unwillingness to listen to one another".

later on, starting from section 1.5). The key point is how to combine the
locality of a set of point-wise differential equations with the global nature
of a purely topological approach. I will show that answering this question
is fundamental to finding the link between classical and quantum physics.
Hence, in a preliminary phase, geometry is selected to be a primary ingredient of my descriptive approach, together with the quantitative power of
differential calculus. On the other hand, as R. Descartes taught us, nature's
laws show up through a clever mixture of both geometrical appearance and
analytic formulas.

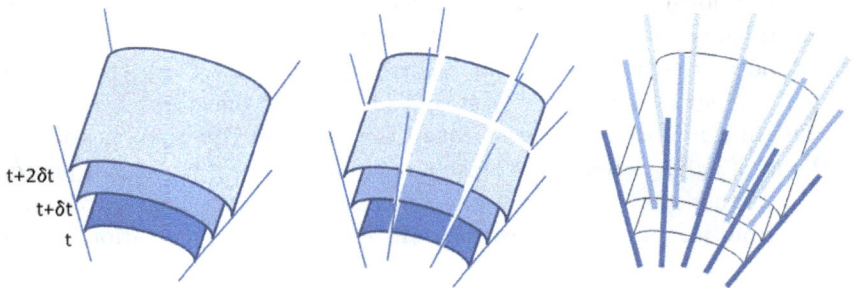

Fig. 1.1 *In accordance with geometrical optics, the different phases of the evolution
of a free solid front (left) agree with the development of its parts (center), up to the
identification with a bundle of rays (right) carrying infinitesimal wave-fronts.*

For instance, as a consequence of my theory, a *dipole wave* can be de-
composed (transverse to the direction of motion) in many solitary sub-
waves, modeled by the same equations and independently following their
own route, that coincides with the global path of the mother wave. In this
way, the entire wave is the superposition of its independent parts, each one
representing another wave. In a limit process, one could see any wave as a
bundle of infinitesimal rays (see figure 1.1). This is true when dealing with
what I call free-waves, where the concepts of "big" or "small" are not ap-
plicable. We may also call these phenomena *propagating waves*, in order to
distinguish them from other, more subtle, electromagnetic manifestations,
that will be analyzed later on.

Despite common belief, very intuitive operations, such as the one of cutting and isolating a piece of front, are strictly forbidden in the framework of Maxwell's equations, and attempting to force their realization is a source of errors. It is however true that the word "photon" finds a more appropriate use in relation to extremely small wave-packets, occurring in dynamical electromagnetic processes associated with matter. Based instead on my viewpoint, photons are just exponents of a large family and are indistinguishable from their companions, whatever the size and intensity. Thus, the infinitesimal solitary waves depicted in the third picture of figure 1.1 are not necessarily real photons (according to physics terminology), and certainly are not the "carriers" of the electromagnetic field. It is my intention to better illustrate this issue in the rest of this chapter.

More in detail, a photon could be hypothetically seen as a tiny bullet, traveling at the speed of light (straight, if undisturbed) and carrying an electromagnetic signal, that is two independent vector fields: the electric and the magnetic. Despite the efforts of R.P. Feynman, who clearly explained how photons can interact with other particles by coding into diagrams the results of these interplays, the internal structure still looks mysterious. This is also valid in force of the fact that photons are not solutions of the whole set of Maxwell's equations[6] (see also footnote 50). In truth, the space of solutions of the Maxwell's system is quite weird and, in contrast to what many celebrated experts may think, does not contain any of the solutions generally considered in most electromagnetic applications. Indeed, as far as "standard" waves are concerned (such as basic spherical waves), the solution space is practically empty, so that it should not be a big surprise to discover that photons cannot be actually modeled. I was very amazed too when discovering that Maxwell's equations are totally inapplicable in the description of what have been considered to be electromagnetic phenomena since the advent of Hertzian waves.

Thus, mathematics severely limits the existence of propagating waves in the Maxwellian universe, so that there is little one can do to save the reputation of the famous equations, at least in the form they are usually employed. Some facts justifying these conclusions are rigorously documented in the book [Funaro(2008)] and my papers. The fulcrum, around which the review process revolves, concerns the divergence of the electric field, hereafter denoted by $\rho = \mathrm{div}\mathbf{E}$. The canonical theory prescribes that $\rho = 0$ in

---

[6]From [Sezginer(1985)]: "The 3-D wave equation does not admit finite-energy nondispersive solutions".

absence of, stationary or moving, electrically charged bodies. So it should be, in particular, within regions touched by pure electromagnetic emissions, far enough away from the sources that generated them. Although we continue to believe that the divergence of the magnetic field (div**B**) must remain zero, the equation $\rho = 0$ is proved instead to be extremely restrictive, resulting in a severe impediment when looking for applications of the standard model equations[7]. In order to make this exposition more fluent, I skip here the details of this important issue, but I collected some facts in appendix B for the reader's sake. Many examples of simple waves of propagating type, that cannot be modeled by Maxwell's equations, are reported in appendix E. The check is a trivial calculus exercise which is strongly suggested to the reader, since it helps in understanding the limits of the classical approach.

I will promise to add further (non-technical) clarifying remarks, during this exposition. Let me just mention that problems often do not emerge from the equation themselves (that are easily studied, as far as some mathematical aspects are concerned), but, more correctly, either from the impossibility of finding suitable initial conditions or from the incompatibility with a large range of boundary constraints. Disappointingly, the analysis is usually carried out by concentrating on the laws of dynamics, while nobody seems to care about the inconsistency with initial or boundary conditions. Thus, the entire apparatus turns out to be a very nice and ingenuous game, having however no pawns to play with.

What I just said above is only in part true. On the one hand, it turns out that the subspace of Maxwellian <u>propagating</u> waves is basically reduced to very few representatives, together with their combinations and variations. On the other hand, the concrete solution space is instead extremely rich, at the point that Maxwellian electromagnetic waves actually fill all the surroundings. These solutions are of a type usually not considered in the mainstream technology[8]. There are however fields in which

---

[7]From [Lehnert and Roy(1998)], p.5: "The idea of introducing a nonzero electric field divergence *in vacuo* originates from some unpublished speculations in the 1960's, in an attempt to explain particles with a rest mass as various eigenmodes of 'self-confined' electromagnetic radiation, somewhat in analogy with magnetic plasma confinement in thermonuclear fusion research".

[8]From [Irvine and Bouwmeester(2008)]: "Maxwell's equations allow for curious solutions characterized by the property that all electric and magnetic field lines are closed loops with any two electric (or magnetic) field lines linked. These little-known solutions ...". The paper then examines the possibility of experimental realizations using knotted beams of light in the framework of laser applications.

such a difference has been noticed. As it will be described in sections 1.7 and 1.8, Maxwell's equations are very well-suited to simulating rotating waves, with a particular preference for those confined in regions with toroidal shape. Numerous examples emphasizing the significance of these solutions are available. Let me just mention for instance the interest they have in the field of *plasma physics* (see, e.g., [Hazeltine and Meiss(2003)]). Anyway, I am going to stop the discussion of this issue here, adding more insight later after more background. To conclude, we find ourselves in a strange situation: Maxwell's model is correct when following families of waves that I consider to be relevant for the description of our universe, but almost irrelevant for basic engineering problems, mostly related to propagating waves; on the contrary, the model is incorrect when applied to solving exactly those practical problems. Unfortunately, to complicate the situation, there are pathological examples displaying an intermediate behavior, as for instance the Hertz dipole solution that will be mentioned at the end of section 1.3.

In the history of mathematical-physics, lots of effort has been made in order to come out with credible models allowing for the inclusion of photons. Let me mention here some publications, where a variety of different models has been proposed, with the aim of extending, for various documented reasons, the theory of electromagnetism: [Born and Infeld(1934)], [Henderson(1980)], [Panarella(1986)], [Hunter and Wadlinger(1989)], [Evans(1994)], [Hunter(1997)], [Lehnert and Roy(1998)], [Benci et al.(1999)], [Munz et al.(1999)], [Yang(2000a)], [Yang(2000b)], [Harmuth, Barrett and Meffert(2001)], [Benci and Fortunato(2002)], [Badiale, Benci and Rolando(2004)], [Benci and Fortunato(2004)], [Coclite and Georgiev(2004)], [Cornille(2004)] [Keller(2005)], [Lo(2006)], [Barrett(2008)], [Donev and Tashkova(2012)], [Arbab(2017)], [Meis(2017)], [Meis and Dahoo(2017)].

Why have I succeeded in pursuing a goal that has been for many years in the must-do list of many ingenuous fellows? Maybe because I did not start from an adaptation of the equations (perhaps, through the introduction of a suitable Lagrangian), but from the field's displacement, and then building up the equations around it. The check that the new model was physically correct came only after.

But, where do we find photons in reality? They are everywhere and, based on my point of view, they are the sole ingredient of our universe. Before facing such an extreme statement, let us stay for a while on more standard ground. According to the common version, photons effectively fill up our environment. They can be "absorbed" by matter that is able to modify their frequency profiles and give them back as the carrier of a new signal. When for instance sunlight, made of an impressive amount of photons carrying an extended range of frequencies, illuminates a material surface, some of these photons become part of the body and contribute to raising its temperature or some alternative form of energy state. Some others may be "reflected" back (immediately or in a short time), reaching our eyes or other instruments. One of the results of this filtering procedure is, for instance, the possibility to distinguish objects of different colors[9]. Similarly, by providing energy to bodies, one may bring them to spontaneously emit photons, as it happens for example in the filament of a lamp. Not all the photons are visible however since their frequencies have an impressively wide range, arriving approximately at 10^{23} Hertz, in the case of the *gamma rays* released by atomic nuclei. So far, I have mentioned several times the term "frequency". This is a very central concept in my theory, since everything started when I began to formalize the idea of an everlasting oscillating behavior of matter at various regimes (see section 3.8). For the moment let us leave the concept of frequency indefinite, although I will use it in intuitive form.

It is important to observe that photons are present at very different dimensional scales. They are in the constituents of nuclei; they appear in atomic and molecular interactions; they are part of biological processes (see section 3.7). All these levels share common properties, but, because of the extreme fragmentation of competence characterizing nowadays the world of scientific research, very little has been done in favor of a unifying interpretation[10]. My aim is to show that connections can be made in a very natural way, once the assumptions of my theory have been understood.

I consider it therefore inadmissible that photons, the quanta of light, such essential ingredients of our universe, cannot have a more proper characterization. An analogous sense of incompleteness has certainly bothered

---

[9]From the introduction in [Guinier(1984)]: "So contemplation of the blue sky has enabled the physicists to say that the colour gives us daily proof of the discontinuous structure of the scattering matter: blue light is more strongly scattered than red because its wavelength is closed to the dimensions of a molecule. If atmospheric air were a continuous medium — in other words, homogeneous at any submicroscopic level however fine — the sky would be black and we should be able to see the stars in full daylight as cosmonauts do outside our atmosphere".

[10]From [Bokulich(2008)], p.11: "What begins to emerge from these defenses of theoretical pluralism is a view of science as 'radically fractured', with distinct scientific communities whose respective members have difficulty communicating with each other".

many scientists in the past. Nevertheless, things developed in a different manner. Indeed Maxwell's equation are not able to include photons within the possible solutions and the various efforts to modify the model turned out to be uncompetitive, if one also takes into account the numerous meaningful properties satisfied by the original set of equations. These defeats were also sources of pernicious ideas, that gradually led to the acceptance that a description does not exist at all, and photons should be taken as they are supposed to be: quanta of light and nothing more. So, photons became basic and unexplainable point-wise particles, carrying frequency and *spin* (in a not specified fashion). This was the beginning of a new era, excellently described by quantum mechanics, where the objects involved cannot be explicitly specified under a certain scale, but reliable predictions can be made on a statistical basis. The assimilation process of such a cultural revolution has been long and troubled[11], but the pill seems now to have been definitively swallowed. A naive example of this paradigm is the description of an atom as a nucleus surrounded by a "cloud" of electrons, according to a certain computable probability distribution. For extended regions containing electromagnetic radiation, one can also introduce the idea of *photon energy density*, that in many circumstances is an effective tool to recover both qualitative and quantitative information, without the need to specify in an exact way what is happening (which turns out to be an impossible request, according to the spirit of quantum mechanics).

Let me honestly say that I have no objections at all regarding the above approach. However, I have always considered the quantum world only as a palliative, a good starting point to proceed with scientific investigations or to get technical achievements, but still waiting for a definitive statement of the basic postulates. This is definitely not true in other contexts, where the quantum way of thinking has been considerably developed beyond the imaginary, becoming a subject for philosophers, sometimes with aberrations that are in contrast to any logical and realistic justification. The only chance to get out of this impasse is to have the courage to blame Maxwell's model, discard the system and revise the rules of electrodynamics in order to include photons in the solution space. I will show how to achieve this scope painlessly.

---

[11]From [Feynman(1963)], 37-7: "Yes! physics has given up. We do not know how to predict what could happen in a given circumstance, and we believe now that is impossible, that the only thing can be predicted is the probability of different events. It must be recognized that this is retrenchment in our earlier ideal of understanding nature. It may be a backward step, but no one has seen a way to avoid it".

1.2 Photon's structure

The photon-type wave is expected to be contained in a cylinder shifting at the speed of light along the direction of its axis. Hence, it has a longitudinal extension along which the electromagnetic fields are modulated in intensity through a one-dimensional function, that in the simplest case may be a single sinusoid period. Moreover, the wave also displays transverse extension, so that a photon is not point-wise. The information lies on parallel slices, orthogonal to the direction of propagation, solidly packed one after the other (see figure 1.2). These flat sections will be called *wave-fronts* and the electromagnetic signal is written on them. As a matter of fact, each front contains the two independent (and orthogonal) vector fields **E** and **B**, representing respectively the electric and the magnetic impulses. The way fields are defined on each front may be quite arbitrary, provided some mathematical hypotheses of global regularity are satisfied.

Another requirement, partly in agreement with the Maxwellian theory, is that the intensity of **E** must be c times the intensity of **B**, where c is the speed of light. This condition is technically referred to as the *vacuum impedance*. We can reasonably impose that the intensity of the electromagnetic field tends to zero when approaching the boundary of the photon. This is aimed at ensuring continuity with the exterior, where no signal is present. As a matter of fact, I intend to solve the modeling equations in the whole 3D space and, for this reason, demanding the smoothness of the solution is an essential request. To avoid the creation of *magnetic monopoles* (see also section 2.1), field **B** should have zero divergence (equation (4) in appendix A). This means that, for any two-dimensional sub-region taken inside a single wave-front, the flux due to the incoming vectors is equal to that of the outgoing ones, i.e., the magnetic field does not display *sources* or *sinks* inside the photon. In other words, because of the boundedness of the object, the magnetic lines of force are compelled to be closed curves.

That's all! Except for the above hypotheses we do not necessitate further ingredients. In practice, the intensity and the orientation (otherwise called *polarization*) of the couple $(\mathbf{E}, \mathbf{B})$ inside each front is not subjected to heavy restrictions. The parallel slices can bring along a different signal and, during the evolution, they do not interfere with each other. The whole set shifts totally unperturbed. We can associate to this movement a velocity field **V** parallel to the axis of propagation and having constant magnitude equal to the speed of light: $|\mathbf{V}| = c$. Of course, the global energy of the packet is preserved during evolution. This is also in full agreement with the fact that the flow of energy has the same direction as the velocity propagation field $\mathbf{V}$. Mathematically, such a property can be checked by computing the so called *Umov–Poynting vector* (proportional to the energy flux), given by the cross product $\mathbf{E} \times \mathbf{B}$, and by recalling that the triplet $(\mathbf{E}, \mathbf{B}, \mathbf{V})$ is orthogonal. This triplet may seem a bit "rigid". Undeniably, there is some redundancy also coming from the condition: $|\mathbf{E}| = |c\mathbf{B}|$. We do not have to forget that I am dealing with the most basic structure; the set of modeling equations will be however furnished with the right amount of degrees of freedom to face more interesting configurations, made available when accelerations are acting on the system (see section 1.7).

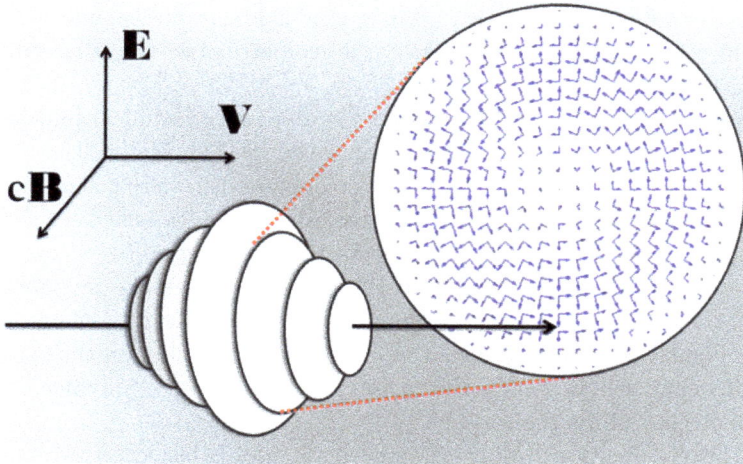

Fig. 1.2 *Shape of a photon according to my model. There are infinite flat independent fronts, each one carrying an electromagnetic signal. The whole structure shifts at the speed of light along the direction determined by* $\mathbf{V}$. *On each front, the magnetic field* $\mathbf{B}$ *follows closed loops and the electric field* $\mathbf{E}$ *is orthogonal to it with* $|\mathbf{E}| = |c\mathbf{B}|$. *The triplet* $(\mathbf{E}, \mathbf{B}, \mathbf{V})$ *is right-handed. The intensity of the electromagnetic fields decays by approaching the photon's boundary. This is also true in the neighborhood of some point located at the interior, in order to allow for the continuity of both* $\mathbf{E}$ *and* $\mathbf{B}$. *For an analytic expression, see for instance (49) in appendix E.*

Let me now try to explain why Maxwell's equations are not suitable for the description of a photon structured as the above one. In addition to a couple of time-evolution equations ((1) and (2) in appendix A), Maxwell's model in empty space implies that both the divergences of $\mathbf{E}$ and $\mathbf{B}$ must vanish. This additional requirement, more related to geometrical properties rather than to dynamical issues, is the real source of trouble. Indeed, it is not difficult to realize that the conditions $\mathbf{E} \perp \mathbf{B}$, $\mathrm{div}\mathbf{E} = 0$, $\mathrm{div}\mathbf{B} = 0$, cannot all hold together. Inevitably, in most cases, the surplus of restrictions brings the trivial solution: $\mathbf{E} = 0$ and $\mathbf{B} = 0$. Thus, it is possible to impose $\mathrm{div}\mathbf{B} = 0$, but it is necessary to drop the condition on the divergence of $\mathbf{E}$, if we do not want the problem to be over-determined. My new model equations, introduced in appendix C, have been actually designed in order to handle the possibility that $\rho = \mathrm{div}\mathbf{E} \neq 0$, while Maxwell's version is more proper when both the electric and magnetic fields are independently organized in closed loops (see footnote 8), a geometrical setting that is rarely associated with propagating waves.

At this point of the exposition, let me give more technical information about Maxwell's model. When first conceived by J.C. Maxwell in the late nineteenth century, the equations of electromagnetism were not in the form they are commonly used nowadays. Successive arrangements were actually made by several authors. However, from the very beginning, a few astonishing characterizations pointed out the potentialities of the model. It has to be remarked that, before the experiments of H.R. Hertz, a few years after the equations were introduced, nobody would have bet on the existence of electromagnetic waves[12]. Hertz himself was not very convinced of the possible utility of his realizations.

As briefly mentioned above, Maxwell's model is based on two coupled dynamical (i.e., time-dependent) equations. The first one is derived from *Ampère's law*. The second one is *Faraday's law* of induction. Combined together, the equations explain for example how a variable electric field may generate a magnetic field, and vice versa. This property is used for instance to build "transformers", that, through coils and ferromagnetic cores, allow for the conversion of voltages or currents. Maxwell's great intuition was

---

[12]In [Maxwell Commemorative Booklet(1999)], F.J. Dyson writes: "The idea that the primary constituents of the universe are fields did not come easily to the physicists of Maxwell's generation. Fields are an abstract concept, far removed from the familiar world of things and forces".

that such a process could be directly made possible by using the sole fields, without the help of macroscopic tools. The reader, however, has to pay attention to a misleading interpretation, also reported in some books (see, e.g., figure 5.2 in [Haken(1981)]), that assumes the evolving fields to behave as chained loops of alternate electric and magnetic fields, displaying some phase difference. This explanation is wrong (see also footnote 4). According to the equations, the <u>traveling</u> fields display the same phase, by meaning that they attain maximum (or minimum) values together.

There are basically two ways of interpreting the solution associated with a propagating wave. From one side, we can follow the evolution standing at a given point and seeing the wave passing by. Alternatively, one can follow the vectors **E** and **B** during their movement along the so called *characteristic lines*, individuated by the tangential vector field **V**. This observation suggests distinguishing between two concepts. The first one deals with the kind of information to be transported, corresponding to the couples (**E, B**). The second one, more recondite, concerns the way the information is transported (in this case a constant shift along straight characteristic lines).

A small ripple produced in the calm water of a long thin basin, takes its message from one extreme to the other. However, in general, there is no net movement of water in the horizontal direction, but, at each point, only a momentary vertical oscillation is registered. Energy is then transferred without actually shifting masses. Of course, here I am not saying anything new; however it is better to clarify things at the very beginning. This training is necessary to prepare the ground for more insidious situations; as for instance the astonishing claim made in section 3.4 that electrons do not need to circulate in a conductor to produce current[13].

As previously anticipated, there are two additional relations, establishing conditions on the divergence of the fields. Roughly speaking, if we are in a totally empty space, one usually constrains the divergence of the electric field to be zero (absence of electric charges), as well as that of the magnetic field (absence of magnetic monopoles). In the end, one has to handle six unknowns (each field has three components) and eight equations (two time-dependent vector equations, plus two divergence scalar conditions). Although there is some (well hidden) redundancy, the fact that

---

[13]From [Feather(1968)], p.5: "We naturally wish to describe an electric current in a conductor as a flow of electrons through its substance, but detailed enquiry convinces us that any picture that we may make of this process in classical terms is fraught with inconsistencies".

there are more equations than unknowns is not a good sign. With little manipulation, it is possible to decouple the two fields $\mathbf{E}$ and $\mathbf{B}$ by recovering two second-order differential equations in vector form (see (6) in appendix A), similar to the scalar *wave equation* describing for instance acoustic phenomena. This was for Maxwell a decisive step towards the validation of his model. In particular, some constants, experimentally measured for Ampère's and Faraday's laws, allowed to compute the velocity of propagation of an electromagnetic pulse, that, with an immense pleasure for the discoverer (I suppose), turned out to be equal to the speed of light (one has: $c = 1/\sqrt{\mu_0 \epsilon_0}$, where μ_0 is the *vacuum permeability* and ϵ_0 is the *vacuum permittivity*).

The study of second-order wave equations is a leading subject in most texts of mathematical physics. Nevertheless, the passage from the vector wave equation to the scalar one by getting rid of negligible components, often performed to simplify the computations, is far from being innocuous. It is indeed a source of errors and misinterpretations that have (negatively, according to my opinion) influenced the successive study of Maxwell's equations (more detailed motivations are found in [Funaro(2008)]). The effect of altering the equations often brings in minor changes in the solution at a local level[14], but irreparably deforms its global aspect, strongly influencing the topology of the wave-fronts (see appendix E). We find ourselves in an embarrassing situation. On one side, due to the divergence constraints and the orthogonality of $\mathbf{E}$ and $\mathbf{B}$, there are practically very few solutions of the two wave-equations ensemble. This makes the model not acceptable from the mathematical viewpoint. On the other side, by retouching a little bit the solutions and introducing some approximation, one ends up with reasonable by-products, useful in a wide range of applications, offering an excellent tool for engineers.

In conclusion, Maxwell's model admits almost no solutions of a propagating type, but well inspired changes may disclose a universe of reliable information about the behavior of electromagnetic waves. We must however face the fact that the original model has no real meaning, without introducing some approximate steps in the analysis. Nevertheless, the model

---

[14]Just to quote an example, from [Kline and Kay(1965)], p.60, we can read: "Since solving Maxwell's equations reduces in many instances to solving the wave equation, it is apparent that the vector theory can be reformulated for scalar functions u satisfying the wave equation". The successive achievements are physically correct but, from the very first passage, the link with Maxwell's equation, whose peculiarity is to work with vector entities, is lost.

was considered to be trustworthy because it allowed the prediction of many phenomena observable in nature. Thus, it seems there are no problems in the end, therefore why am I complaining? I disagree with the statement that Maxwell's equations represent the major reference point in classical electromagnetism. Having a drastically reduced space of solutions, they are unable to predict the existence of photons, for example. The reason why photons cannot be modeled is not because they are out of the limit of the equations (and thus do not belong to classical physics), but because these limits are too strict.

I can recap by saying that a photon-like emission is a compacted bunch of orthogonal triplets. The electric and magnetic fields are transversally placed with respect to the direction of motion, which is indicated by the velocity field $\mathbf{V}$. We know that photons can travel long distances with negligible deformation. It is the case for instance of those emanated by far away stars. They reach our Earth after a journey at a crazy speed, showing no signs of dispersion. Their structure has resisted this long trip, so that, from the frequency spectrum they carry we can recognize nuclear and chemical reactions in the places of origin.

The role of photons in the micro-world is well recognized and we will have time during this exposition to analyze it in full. However, the way photons have been defined here does not imply that they must be small. As a matter of fact, at the beginning of section 1.1, I specified that there is no distinction between a photon and a generic electromagnetic wave. Does this entail that there could be photons of any size? My answer is going to be ambiguous. Free photons are usually emitted by atoms, through a procedure that will be clarified in chapter three. Therefore they are indeed very small. Larger photons could be emitted by some manmade tool, such as an antenna. Note that establishing point-to-point communications without significant dissipative effects is a priority feature in many applications Engineers know that antennas are mainly omnidirectional and that it is extremely difficult to pilot a signal towards a prescribed target. From the theoretical viewpoint, the mechanism of emission of an antenna device is still an open question. Also mysterious is the reason why the output is so diffused. In a center-fed dipole the signal originates from the feeding source, but the effective emission requires the whole resonant apparatus[15]. A pos-

---

[15]From [McDonald(2014)]: "As Sommerfeld reminds us: conductors are nonconductors of energy. Rather than thinking of the conductors of an antenna as source of radiation, it may be better to consider them as a kind of inverse wave guide, which 'tells' the radiation where not to go. [...] This reinforces the view that the conductors of the antenna serve

sible conclusion is that big photons may theoretically exist but their perfect realization seems out of reach because of impediments due to the instrumentation used. Let me just mention a technical problem. As I specified before, free emission is characterized by being represented by orthogonal triplets $(\mathbf{E}, \mathbf{B}, \mathbf{V})$, with $|\mathbf{E}| = |c\mathbf{B}|$ and $|\mathbf{V}| = c$. For reasons inherent to the properties (up to atomic scale) of the conducting material used for the emitter, this initial assessment is not verified. In particular, the electric field tends to have a longitudinal component in the direction of propagation[16]. A wave not subjected to further perturbations tends to restore the orthogonality of $(\mathbf{E}, \mathbf{B}, \mathbf{V})$, but the imprinting at the origin is decisive for addressing its orientation. Technology may surely provide new ideas to handle the asperities, and come out with very competitive devices, as far as directivity is concerned. Anyway, it is important first of all to remove ambiguities and this is the scope of these pages. I can be more precise as the whole final set of model equations will be introduced.

We are starting here to be confronted with questions that are at the border between classical and quantum theories, which is exactly the argument I would like to study in these notes. By looking at figure 1.1, we may be tempted to argue that a wave is the sum of its "photons". Noting instead that photons are waves themselves and can have potentially any size, has serious methodological implications. In fact, the role of photons as "carriers" of the electromagnetic field becomes meaningless. On the other hand, wave-fronts produced by an antenna are usually depicted as solid smooth surfaces; this is just a rough mathematical simplification of an intricate phenomenon of photon emission, originating from the device itself. In this case, a set of model equations can only provide reliable indication of averaged behavior, but to know more about the real constitution and the way the emission initially takes place, it is first necessary to carry out investigations on the structure of conductors. As the reader can notice, there are several concepts fused (and confused) when talking about electromagnetic waves. My wish is to bring order and clarity. Thus, let us continue to analyze the most simple phenomena before facing more complex ones.

---

to redirect the waves 'created' at the central feed point, rather than to create the wave themselves".

[16]From [Keller(2005)], p.42: "The scalar and longitudinal photons only exist in the near-field zone of matter, and only as long as the particle source emitting the transverse photons is electrodynamically active. The longitudinal and scalar photons cannot be detached from the matter field and they play an important role in linking quantum optics and near-field optics together".

1.3 A closer look at the model

The equations simulating electromagnetic phenomena must be energy-preserving *transport equations* of *hyperbolic type*. Differently from other situations (for instance acoustics), the information carried is not of scalar type, i.e., a single quantity such as a density of mass, energy or pressure. In fact, the equations in the electromagnetic case deal with a far more complicated set of variables, representing the two independent vector fields **E** and **B**. As I claimed in the previous section, there is the tendency to simplify the problem by eliminating "negligible" components of the vector fields and reduce the analysis to scalar variables (see footnote 14). Most of the time this might not correspond to a reasonable approximation of the vector equations and people often tend to abuse such a procedure; therefore, I strongly believe that a serious approach must take into account the whole complexity of the setting.

My model makes the distinction between *free-waves*, where the evolution proceeds unperturbed, and *constrained waves*, where for external reasons the evolution of the wave deviates from the natural path. All these processes evolve at the speed of light, which is basically the only velocity we actually find in my construction. In order to fully understand the last statement, it is necessary to embed the analysis of the equations in the habitat of *general relativity*. Before facing this complicated issue, I would prefer to give a certain number of insights that do not necessitate fancy arguments. I will begin by describing the case of free-waves, which is simpler. The general equations ensemble is displayed in the next page together with the main subcases. In particular, free-waves are discussed in appendix C.

Let me first provide a brief description of the revised equations by anticipating some comments. Different from the classical ones, the set of equations is of nonlinear type. This must be expected, since all physical observations converge to a nonlinear description of natural events. Linearization may be helpful to get a rough idea of the facts, but it is deleterious when applied beyond certain limits. The second relevant aspect is that the set of equations operates on three fields $(\mathbf{E}, \mathbf{B}, \mathbf{V})$. Thus, together with the electromagnetic information, the model also describes the way this information has to be transported (see section 1.2). Since there are now more unknowns, there is the need for an extra vector equation, that, in the case of free-waves, turns out to be just a geometrical constraint, establishing the orthogonality of the triplet $(\mathbf{E}, \mathbf{B}, \mathbf{V})$. Such an equation reads as follows: $\mathbf{E} + \mathbf{V} \times \mathbf{B} = 0$.

By defining $\rho = \text{div}\mathbf{E}$, the general set of equations is:

$$\frac{\partial \mathbf{E}}{\partial t} = c^2 \text{curl}\mathbf{B} - \rho\mathbf{V}$$

$$\frac{\partial \mathbf{B}}{\partial t} = -\text{curl}\mathbf{E}$$

$$\text{div}\mathbf{B} = 0$$

$$\rho\left(\frac{\partial \mathbf{V}}{\partial t} + (\mathbf{V}\cdot\nabla)\mathbf{V} + \mu(\mathbf{E}+\mathbf{V}\times\mathbf{B})\right) = -\nabla p$$

$$\frac{\partial p}{\partial t} = \mu\rho(\mathbf{E}\cdot\mathbf{V})$$

where μ is a given constant and c is the speed of light.

A continuity equation easily follows from the first equation:

$$\frac{\partial \rho}{\partial t} = -\text{div}(\rho\mathbf{V})$$

Free-waves are obtained as a special case, i.e.:

$$\frac{\partial \mathbf{E}}{\partial t} = c^2 \text{curl}\mathbf{B} - \rho\mathbf{V}$$

$$\frac{\partial \mathbf{B}}{\partial t} = -\text{curl}\mathbf{E}$$

$$\text{div}\mathbf{B} = 0$$

$$\rho\,(\mathbf{E}+\mathbf{V}\times\mathbf{B}) = 0$$

Maxwell's equations in vacuum are finally obtained for $\rho = 0$:

$$\text{div}\mathbf{E} = 0$$

$$\frac{\partial \mathbf{E}}{\partial t} = c^2 \text{curl}\mathbf{B}$$

$$\frac{\partial \mathbf{B}}{\partial t} = -\text{curl}\mathbf{E}$$

$$\text{div}\mathbf{B} = 0$$

The substantial achievement of the new model is the ability to get rid of the zero divergence condition on the electric field (the primary source of impediment) and modifying accordingly the Ampère law. There is now a "current term" in the Ampère equation, given by: $(\mathrm{div}\mathbf{E})\mathbf{V}$ (see (16) in appendix C). This is not an effective current due to an external device, but rather something flowing with the wave itself with a certain density $\rho = \mathrm{div}\mathbf{E}$ and velocity $\mathbf{V}$ equal in magnitude to the speed of light. Like currents circulating in a superconductive wire, photons travel in a straight line at speed c, preserving their bounds. Without adding further hypotheses, a *continuity equation* for the scalar ρ can be easily obtained by differentiating the modified Ampère's law (see (22)).

I will provide better explanations of the entire construction later in this chapter, when dealing with constrained waves, that come from generalizing the present setting. This extension also includes a new scalar potential p (see previous page). For the sake of completeness, let me inform the reader more concerned with technical details that a *Lagrangian* is available and that the equations are naturally connected to the divergence of the electro-magnetic *stress tensor* (see appendix D). With respect to these parameters the new model is as good as Maxwell's. The substantial gain is that it admits a larger space of solutions.

Whatever their shape is, free-waves realize the Galilean concept of a "body" traveling straightly at constant uniform speed. The family of photons, introduced in the previous section, consists of free-waves with parallel propagating fronts. An easy and typical example of free-wave is the plane wave, where all the fronts are actually entire planes and the fields defined on them are constant everywhere. Such an entity does not exist in nature, since its transverse extension as well as its energy are infinite. By the way, this is the only solution (up to changes of intensity and polarization of the fields inside each front), based on parallel flat fronts, that is admitted by the set of Maxwell's equations. Other Maxwellian propagating waves can be found by assuming that the fronts are not planar, but still unbounded and carrying infinite energy. If we opt for bounded fronts, chances to find solutions are very few. Correspondingly, one also has to assume that the Umov–Poynting vector is not lined up with the direction of propagation of the fronts, which is an unphysical situation.

Free-wave fronts can display a multitude of different shapes. A survey of possible solutions satisfying the new set of equations is found in appendix E. An infinite number of other options are possible. In a pure spherical wave, the fronts are the skins of concentric spheres. This means that the field **V** is still constant, but of radial type. On the surface of each sphere we find the electromagnetic signal. More precisely, at each point, **E** and **B** (still orthogonal) belong to the local tangent plane to the spherical surface. Again, we are (almost) free to choose intensity and polarization on each front as we prefer (as far as we respect the div**B** = 0 condition), and, during the evolution, there is no interference between the fronts. Nevertheless, to guarantee the preservation of energy, the intensity of the electromagnetic field has to fade, since the same original signal is going to be distributed on surfaces whose area is growing with time. In this situation the information is diffused everywhere, and arrives at any single destination with reduced strength. The model automatically takes care of these circumstances. The further condition div**E** = 0 is incompatible with this setting, saying that Maxwellian spherical wave-fronts are forbidden.

Let me stress once again that there is no direct relation between the signal transported and the way it is transported (see figure 1.3). In a spherical evolution of the fronts, **E** and **B** do not necessarily satisfy any symmetry constraint. They constitute the message to be transmitted (any message), that may be transversely arbitrary and suitably modulated longitudinally. The adjective "spherical" is only associated with the shape of the fronts and the way they move, which is dictated by the vector field **V**. Commonly, when displaying signals, engineers do not make these distinctions. Indeed, these doubts do no touch professionals in specialized areas; those are too confident of their paraphernalia to look into such sophisticated questions. That is one of the reasons why certain inconsistencies never came to light. This is especially true when dealing with well tamed solutions, such as the *Hertzian dipole*, that will be briefly handled later in this section.

Fig. 1.3 *These ripples in water, taken at fixed instant, are the information to be transported. They can be erroneously confused with the first picture of figure 1.1, showing the evolution of a single front in time, underlining the way a certain piece of information is transported. Oscillatory phenomena, such as this one, test our capacity to discern between space and time, and between what and how.*

There are many other intermediate situations that can be faced with the following formula: fronts developing according to a stationary constant velocity field $\mathbf{V}$, and electromagnetic information imprinted on the tangent planes to the fronts, in order to form orthogonal triplets marching in empty space. At this point, let me remark that, in conceiving the model equations, my attention was concentrated on the fact that I wanted the solutions to behave in a certain manner. Therefore, the whole machinery was built around the a priori knowledge of exact solutions. Of course, once the model is written, one can start thinking about the characterization of the whole space of solutions. I have some results in this direction, although I did not pay too much time on specific theoretical issues. In particular, I have no idea how a general imposition of the initial conditions may influence the future evolution of a wave. For example, one could imagine initial fronts where equation $\mathbf{E} + \mathbf{V} \times \mathbf{B} = 0$ is not fulfilled (see the comments at the end of section 1.2, about fields emitted by an antenna).

Now, it should be clear that the laws ruling the evolution of the fronts are not connected with the signal transmitted. Based on *geometrical optics*, in an amount of time δt each front covers, in the direction point-wise orthogonal to its surface, a distance $\delta x = c\,\delta t$, c being the speed of light. In vector form, such a relation becomes: $\delta \mathbf{x} = \delta t \mathbf{V}$. Thus, each point of a front can evolve independently from the others, following a straight trajectory comparable to a light-ray (see figure 1.1). My model equations are actually in agreement with such a way of evolving vector waves, and this is quite an amazing property. In fact, I am dealing with partial differential equations that combine time and space derivatives of the components of electric and magnetic fields. I assume that, at some initial time, $\mathbf{E}$ and $\mathbf{B}$ are set up in order to be compatible with the successive evolution. This means that the electric and magnetic vectors must be mutually orthogonal, and placed on tangent planes of surfaces that will become the evolving fronts. Once this is done, the equations allow for the development of the global wave according to geometrical rules. Surprisingly, if the shapes of the fronts were initially the same, but with modified electromagnetic information, the subsequent evolution of the fronts would not change, despite the fact that the equations are built on the specific values attained by the carried fields. In this fashion, the model can reproduce the behavior of a free photon independently from the message carried by it. This is a first example of what we can call *globalization*, where the support of infinite point-wise unconscious contributions, following a specified law, produces an all-embracing effect that does not rely upon the quantitative nature of the initial population, but only on its topological set up. Moreover, the equations preserve physical entities such as energy, so that, as the case of spherical wave-fronts requires, the intensity of the fields decays when the fronts develop. I am sure that somebody, more expert than me, may find interesting explanations for the new model equations within the framework of differential geometry. Other claims reported here are not supported by a scrupulous analysis. I understand that some inflexible pure mathematician could be scandalized by the lack of rigorous results. Nevertheless, I always thought that it is possible to breathe even if one is not able to prove that air satisfies the 3D fluid dynamics equations.

I consider it a severe handicap that spherical wave-fronts cannot be build in the Maxwellian context (see [Funaro(2010)]). With such a poor space of solutions, the classical model is certainly unable to deal with the immense and variegated universe of electromagnetic phenomena. This is one of the reasons I felt that a revision could not be further postponed. The literature is full of rough efforts to adjust and combine plane waves in order to get other possible propagating solutions, using the linearity of Maxwell's equations. Some of these approaches are mathematically questionable, thus showing how intuitions not well justified can lead to unrealistic conclusions.

In textbooks, plane waves are described by (scalar or vector) expressions containing the multiplicative term $\exp(i(\mathbf{k} \cdot \mathbf{x} - \omega t))$, with $\mathbf{k}$ denoting the direction of movement. Such a writing does not represent a "propagating" entity, but something that exists for all t, hence also in the future, i.e., in places where the message is still not supposed to be. Of course, one can try to restrict the support in order to work with solitary shifting pulses, although troubles emerge from the imposition of boundary conditions (as far as the scalar variable $\xi = \mathbf{k} \cdot \mathbf{x} - \omega t$ is concerned), since the evolution speed of each *Fourier mode* depends on the magnitude of $\mathbf{k}$, resulting in a distortion of the original packet. Moreover, all those concepts, such as *phase velocity* and *group velocity*, well-suited for the analysis of periodic structures, become meaningless when studying a single pulse. Again, problems originate from an incorrect identification of what is transported and the way it is transported (see the caption of figure 1.3). With these prerequisites, connections with geometrical optics are imprecise and supported by weak justifications[17].

---

[17]From [Kline and Kay(1965)], p.14: "All one can really deduce from the study of plane waves is that they obey some of the laws of geometrical optics but they do not suffice to derive geometrical optics from Maxwell's equations". Unfortunately, I claim that there is no way at all to get geometrical optics from Maxwell's equations. Results at local level

After suitable "normalization", R.P. Feynman sets the foundations of *Quantum Electro Dynamics* by linking photons to classical Maxwell's equations[18]. He makes use of mystifying passages, generating self-confidence in borderline questions that should instead necessitate rigorous mathematical clarification. There is also a strong tendency to avoid the distinction between time and space variables, so that a plane wave ends up to be a promiscuous object, conveniently reduced to a moving point when required by the context. The result is a picturesque representation where a photon is sketched by a little snake. With this I do not want to cast a shadow on the universally recognized success and consistency of *QED*, though the very initial steps are embarrassingly approximate.

Solutions of Maxwell's equations are often introduced as expansion series in terms of well-known basis functions. A consolidated technique is to integrate with respect to $\mathbf{k}$ a family of plane waves of the type $\mathbf{A_k} \exp\left(i(\mathbf{k} \cdot \mathbf{x} - \omega t)\right)$, for suitable vector coefficients $\mathbf{A_k}$ with $\mathbf{A_k} \cdot \mathbf{k} = 0$ (see for instance [Clemmow(1966)]). In this way, the divergence of the resulting vector field turns out to be equal to zero. Unfortunately, excluding a few exceptions, the lines of force of the final field are not tangential to the evolving fronts, so missing the links with geometrical optics and the possibility of including photons in the analysis. It is not possible to have a cake and eat it too: either one preserves the correct advancing of the fronts or realizes the divergence free condition[19]. Ironically, the above mentioned approach is the preferred one in classical optics[20,21]. The confusion may

---

are achievable but a global control of the fronts cannot be accomplished.

[18] From [Feynman(1998)], p.4: "Thus, in general, a photon may be represented as a solution of the classical Maxwell equations if properly normalized. Although many forms of expressions are possible it is most convenient to describe the electromagnetic field in terms of plane waves".

[19] For example, in the expression $(1,0)\sin(y - \omega t) - (0,2)\sin(x - \omega t)$, the modulated 2D signal $(1,-2)$ is transported along directions parallel to the vector $(1,1)$. The divergence of the field is zero but the signal is not tangent to the fronts, which are straight lines orthogonal to $(1,1)$.

[20] From [Hecht(2002)], p.27: "... first, physically, sinusoidal waves can be generated relatively simply by using some form of harmonic oscillator; second, **any three-dimensional wave can be expressed as a combination of plane waves**, each having a distinct amplitude and propagation direction".

[21] From [Stone(1963)], p.478: "By suitably assigning amplitudes, polarizations, and phases to the plane-wave components, one can construct a spherical wave or any other

be once more attributed to the unmarked distinction between the front surfaces and the envelopment of the lines of force of the electric field.

In other contexts, people tend to forget that planes are huge cumbersome objects. Maxwellian plane wave-fronts carry a constant electromagnetic signal that spreads out; they cannot be cut without generating discontinuities and the information written on them cannot be concentrated in some areas, without affecting the divergence of the fields **E** and **B**, which in vacuum is expected to be zero everywhere. It is unclear then how these unmanageable things can be used to "approximate" curved fronts, both regarding their geometry and the signal written on them[22]. In many numerical codes, fronts are discretized through a sequence of piece-wise flat surfaces, so neglecting the fact that a plane wave has infinite spatial extension. In such a procedure it is not guaranteed that the conditions on the divergence are maintained passing to the limit (and this is usually false if the approximated front is a simply connected region). The process recalls that of approaching uniformly a 1D differentiable function using piece-wise constant functions. The approximating functions have zero derivative almost everywhere, but this property is certainly not transferred to the limit. This last example refers to the stationary case, but the situation does not improve if one also takes into account time evolution.

Nevertheless, the task of making evident the little crimes perpetrated in more than a century is not that easy. Like an unprepared student who reaches the right answer in his homework, after a sequence of mistakes canceling each other out, the by-product of the incorrect rearrangement of plane and spherical waves ends up fitting real-life experiences quite well, within acceptable limits[23]. In numerical experiments concerning the scattering of a wave by an obstacle, a common mistake is to follow the dynamical equations (1) and (2) without worrying about the divergence conditions,

---

type of radiation field arising in the classical theory".

[22] From [Keller(2005)], p.32: "Photon wave mechanics is based on the introduction of a photon wave function, and the establishment of a dynamical equation which can describe the space-time evolution of the probability amplitude. The wave function of a spatially localized photon must have wave packet character, and in classical electrodynamics free-space electromagnetic wave packets can be formed by superposition of monochromatic plane waves". This interpretation is inspired by mono-dimensional examples. As one tries to argue in 3D, contradictions soon emerge: how can plane waves be combined to form a spatially localized wave-packet?

[23] From [Clemmow(1966)], p.20: "The concept of an unbounded plane wave is, of course, an idealization. Nevertheless, in theoretical work an unbounded homogeneous plane wave in a lossless medium is a convenient fiction which can be physically acceptable".

since these have been already imposed on the incoming wave (normally a plane wave, what else?). Irreparably, for certain types of boundary conditions, during and after the impact with the obstacle at least one of the two fields shows a divergence different from zero in its neighborhood. Such a deviation is usually not tested, though the error introduced does not reduce by refining the computational accuracy. In this way incorrectness proliferates, nourishing the initial misunderstandings surrounding the legendary equations. In spite of that, the final results agree quite well with those of laboratory experiments.

The constant technological progress in radio communications seems to testify in favor of Maxwell's equations. I continue to claim instead that a model with such a reduced space of solutions has no reason to survive and one of the consequences of having accepted it, is the neat division between the classical and quantum worlds. It is also to be remembered that my modification of the original equations is not drastic (just a small alteration). Theoretically, it can be physically accepted with no major conceptual efforts. Practically, it provides the world of applications with an extremely large space of effective solutions.

There is an explicit spherical-type solution, called the *infinitesimal dipole solution* found by H. Hertz himself, solving analytically the whole set of Maxwell's equations in vacuum (see (54) in appendix E). Its corresponding wave-fronts (the envelope surfaces of the couples $(\mathbf{E}, \mathbf{B})$) are toroidal (not spherical) with a banana-like section and, as actually prescribed by the conditions $\mathrm{div}\mathbf{E} = 0$ and $\mathrm{div}\mathbf{B} = 0$, the fields form closed loops (see figure 4.3). The Umov–Poynting vectors are not orthogonal to spherical surfaces, so that the rules of geometrical optics are not fulfilled. The way energy is transferred in this process, for instance in the study of an antenna device, is still (after more than one hundred years) reason for debate among technicians. Approximations of the dipole solution are available; they do not solve the Maxwell's system (basically because they forget to impose the $\mathrm{div}\mathbf{E} = 0$ constraint), but they better agree with an ideal spherical wave.

The Hertzian solution is considered to be a pedagogical example illustrating the goodness of the Maxwellian theory. From my viewpoint its existence is a misfortune. For the reasons detailed in [Funaro(2008)], that for simplicity I do not wish to report here, the behavior of the Hertz fields is, in its generality, quite unphysical, although locally very similar to what is observed in reality. This equivocal situation is a source of misunderstandings when talking with experts. My efforts to explain my version about the propagation of fronts in relation to the Maxwell's model have been very often frustrated by referring to the Hertzian solution as the panacea of all troubles. Plane and Hertzian waves (both non existent entities in real life) seem to be the only weapons in the hands of Maxwell's supporters, but they are heavily employed when repressing any uprising.

Let me summarize what I can actually do with my model. Without introducing any significant physical alteration, I can include in the solution space all those waves, mentioned above, that are well recognized in engineering applications (like some well-behaved approximations of the Hertzian solution). Moreover, I can include solitary waves, such as photons. Therefore, let us not waste more time with the old stuff. We have a new toy now. After appreciating its numerous capabilities, the reader may turn his head back and realize how reductive the previous view was. The next step will be to examine the role in the framework of the theory of relativity. Amazingly, we will also rehabilitate Maxwell's equations; in fact they are correctly posed if we look at them in the proper way.

1.4   Connections with special relativity

A first serious investigation on the properties of light started with the work of A. Einstein, based on preliminary results by H. Lorentz. As a consequence, Maxwell's equations, one of the central issues in Einstein's analysis, gained further consensus. Roughly speaking, with an appropriate rewriting of the Maxwellian model, it is possible to get an "invariant formulation", adaptable to almost any kind of observers. According to such an intrinsic set up, the qualitative features of a given light phenomenon can be properly described, and the same conclusions may be reached from various different observational frameworks[24]. The first achievement in this direction is the proof of the Lorentz invariance of Maxwell's equations. This result contains many of the ingredients of the theory of *special relativity*. The property is checked by applying a time-dependent change of variables (namely, a Lorentz transformation), where the new reference frame moves with constant speed **v**, i.e., it is an *inertial frame* with respect to the one considered to be at rest. It is then discovered that each inertial observer is able to

---

[24]From [Einstein(1905)]: "... Examples of this sort, together with the unsuccessful attempts to discover any motion of the earth relatively to the 'light medium', suggest that the phenomena of electrodynamics as well as of mechanics possess no properties corresponding to the idea of absolute rest. They suggest rather that, as has already been shown to the first order of small quantities, the same laws of electrodynamics and optics will be valid for all frames of reference for which the equations of mechanics hold good".

interpret from his own viewpoint a given electromagnetic emission with the
same universal modeling equations.

The exercise is carried out as follows (see appendix F). With little
computational effort, one discovers modified equations resulting from the
change of variables applied to the Maxwell's model, i.e., by adapting partial
derivatives in relation to the transformation. One has to verify that any
Maxwellian wave described in the rest frame is seen in the moving frame as
a solution of such modified equations. In addition, one discovers that the
modified model equations have the same "appearance" as the ones stated
in the rest frame (compare (65) in appendix F with (12) in appendix A).
Thus, there is a common denominator that links equations and their corresponding solution spaces, independently of **v**. Unfortunately, Einstein was
not aware of the fact that, as far as propagating waves are concerned, the
solution spaces are almost empty. This is a pedagogical example on how
interesting abstract results can be proven without having representatives to
satisfy them. Nevertheless, based on what has been reported in the previous sections, there are other less known waves solving Maxwell's equations,
hence the situation is not so critical (see footnote 8). This kind of solutions
will be extensively used in the following.

Another important aspect is the invariance of the speed of light. In
particular, two observers, one in motion (with constant velocity **v**) with
respect to the other, would deduce that light evolves at the same speed
c, as indirectly recovered by A. Michelson and E. Morley in their famous
experiment, trying to measure the effect of Earth's motion on the speed of
light[25]. Consequently, in disagreement with intuition, velocities do not sum
up linearly; namely one has $|\mathbf{V} + \mathbf{v}| = c$, even if $|\mathbf{V}| = c$ and $\mathbf{v} \neq 0$ (here
the symbol $+$ has to be interpreted in the proper way). In other terms, for
a spherical wave the fronts develop radially all around at the same speed
c, and this remains true also if one looks at the wave, while shifting with
constant motion. Ironically, this argument does not take into account that
spherical fronts of light (and other isomorphic shapes) are not members of
the Maxwellian population. I am sure that, if scientists were conscious of
this fact in the past, they would have soon found a remedy. However, this

---

[25]From [Landau and Lifshitz(1961)], p.3: "Measurements first performed by Michelson
(1881) showed complete lack of dependence of the velocity of light on its direction of
propagation; whereas according to classical mechanics the velocity of light should be
smaller in the direction of the earth's motion than in the opposite direction". The aim
of the experiment was actually to prove the existence of a sort of aether wind; with this
respect the outcome was a failure.

incoherence was not pointed out, probably not being pondered with the necessary thoroughness.

The result of the invariance of the equations of electromagnetism is one of the pillars of the theory of special relativity. Therefore, it is important to verify that Lorentz transformations are compatible with my model too. Not only does this turn out to be true in the case of free-waves (I give the details in appendix F, that imitate those in [Einstein(1905)], i.e., the original Einstein's paper), but there are additional interesting consequences to remark upon. Let me first recall that the two postulates of special relativity are generally stated as follows. The first one says that physical laws are invariant under inertial coordinate transformations of Lorentz type: if an object obeys some mathematical equations in a frame of reference, it will obey the same type of equations in any other frame shifting at constant velocity **v**. This is exactly what happens for the usual equations of electrodynamics. The second postulate says (see above) that light always propagates in empty space with speed c, independently of the state of motion of the emitting body. From this, Lorentz also deduced a nonlinear formula for summing up velocity vectors (see (70)).

Now, when proving the Lorentz invariance of the new set of equations one has to deal with **E** and **B** as in the traditional case. It has to be remembered however that the field **V** also appears in the equations. If the observer is moving with constant velocity **v**, the transformation of **V** in the new reference frame has to be computed with the nonlinear formula for velocities, combining both **v** and **V**, where here the magnitude of **V** is equal to c. The proof of the Lorentz invariance has to take into account this correction, which is exactly what Einstein did when adding the Ampère current term to Maxwell's equations[26]. The only difference is that, in my model, currents are not due to external factors, but they are part of the field description. Hence, my approach is not actually introducing any technical novelty, but only a different way to explicate facts. Finally, in addition to these accomplishments, one needs to check also that the free-wave condition (19) is Lorentz invariant (for example, spherical propagating fronts are

---

[26]In [Einstein(1905)], after proving the invariance of Maxwell's equations including the current source term, the author writes: "... we have the proof that, on the basis of our kinematical principles, the electrodynamic foundation of Lorentz's theory of the electrodynamics of moving bodies is in agreement with the principle of relativity. In addition I may briefly remark that the following important law may easily be deduced from the developed equations: if an electrically charged body is in motion anywhere in space without altering its charge, when regarded from a system of co-ordinates moving with the body, its charge also remains when regarded from the 'stationary' system ...".

represented by free-waves, and this property must turn out to be true in any inertial frame). Hence, it is necessary to work on the condition $\mathbf{E} + \mathbf{V} \times \mathbf{B} = 0$. In order to achieve the result, the nonlinear addition formula for velocities has to be reused (see also [Funaro(2008)], section 2.6). The verification is going to be quite an easy job for an expert reader.

By reversing the order of the passages, one entails the following consequence: if my model is Lorentz invariant (including the equation $\mathbf{E} + \mathbf{V} \times \mathbf{B} = 0$), then the second postulate must hold. Indeed, the fact that a spherical wave is "spherical" for all inertial observers is directly connected to the constancy of the speed of light. In other terms, spherical signals emitted by a point-wise source reach their targets independently of the shifting velocity $\mathbf{v}$ of such a source. The above observations suggest a rethinking of the postulates. Note that Lorentz formulas already establish a link with c.

Observing that my model complies with all those basic rules, was for me a good source of optimism. Once again, let me say that these deductions were made by Einstein without having any concrete example of propagating spherical fronts, because in the Maxwellian context they just do not exist. Let me stress that these fronts have to carry vector signals, therefore, spherical advancing fronts for scalar fields are not acceptable. Let me also stress again that with the adjective "spherical" I mean the geometrical form of the signal "carrier", which has nothing to do with the signal itself. For the reasons partly put forth in section 1.3, the Hertzian dipole is not an acceptable example of spherical wave-fronts propagation (see figure 4.3). Moreover, let me admonish that the *eikonal equation* (the fundamental equation of geometrical optics) describes very properly the way the fronts must develop, but does not provide any insight about the carried message (recall the distinction made in section 1.3).

Since I mentioned the eikonal equation, let me spend a few words about it. This is a nonlinear partial differential problem (see (20) in appendix C), allowing for the determination of time-evolving surface fronts, starting from some initial configuration. According to this approach, the fronts are interpreted as the 2D level sets (analogous to the 1D level lines in 2D cartography) of a function Ψ in 3D, representing a scalar potential. We want each level set to be uniformly distant with respect to the neighboring ones, i.e., we can pass from one set to another by moving all points of an identical distance δx, along the normal direction to the surface. The gradient $\nabla \Psi$ of Ψ is a 3-components vector orthogonal to the 2D level sets. In the stationary case, the eikonal equation amounts to imposing that $\nabla \Psi$ has a constant modulus everywhere it is defined. The constant turns out to be equal to c, that is the speed of light (the velocity of the marching fronts). If we denote by $\mathbf{V}$ such a gradient, we get the velocity field of my model equations. The other way around, we can argue as follows. If $\mathbf{V}$ can be written as the gradient of a scalar function Ψ (which is expected to be a reasonable assumption in the case of free-waves), then the constancy of the modulus of $\mathbf{V}$ brings to the eikonal equation $|\nabla \Psi| = |\mathbf{V}| = c$, and, at the same time, expresses the constancy of the speed of light. The corresponding surface fronts turn out to be orthogonal to the field $\mathbf{V}$. This justifies from another viewpoint the presence of $\mathbf{V}$ in the new modeling equations and establishes a closer link with geometrical optics.

Somehow, the velocity field $\mathbf{V}$ decides the direction of motion of the fronts and forces, at any point, the couple $(\mathbf{E}, \mathbf{B})$ to lay on the tangent plane of each corresponding front. If we do not impart any acceleration to $\mathbf{V}$, the motion will be undisturbed, the fronts proceed according to the rules of optics and the light rays are straight lines (the two typical examples are spherical waves or photons, but the family includes uncountable members). Note that, if we try to solve the new model equations with an initial $\mathbf{V}$ having a modulus different from c, there are no chances of getting a free-wave. The mechanism that docilely drives the electromagnetic fields to envelope the right fronts is missing in Maxwell's approach, with the result that there is no trace of photons in the solution set. As a matter of fact, the Maxwellian solution space originates from the condition $\rho = 0$, that sweeps away the term containing $\mathbf{V}$, breaking any contact with the eikonal equation. I understand that the above presentation is incomplete and intuitive; therefore, a more robust mathematical formalization of the results is expected. Let me conclude this paragraph by remarking that the lack of interconnections between Maxwellian electromagnetism and geometrical optics is well-known. Links are actually very mild and can only be found within the framework of heavy approximations, where, basically, one discovers that a smooth front behaves locally as a piece of plane wave. I already made my comments in section 1.3 about the unreliability of such a procedure (see also footnote 17).

I think that the ideas presented so far can be accepted with no major objections by scientists. In the end, the basic theoretical tools I am using here have been matured for as long as hundreds of years and belong in the background preparation of a good physicist. There are however some approximate passages in the habitual exposition that, although intuitively correct, did not help the full development of the concepts involved. I refer for instance to the ambiguous transition from the vector Maxwell's equations to the scalar eikonal equation, and, once again, to the lack of distinction among the signal carried and the way it is carried. A typical phenomenon, always mentioned in special relativity, is the *Doppler effect*. Its study is mathematically correct, nevertheless the explication of the facts is a bit fuzzy. Let me make some comments about it.

If a periodic process of period λ takes place in a moving frame, then the period registered at the rest frame is going to be greater than λ if the motion is away from the measuring point and less if the motion is towards it. The change of frequency of a moving siren testifies to the validity of this principle. It is said that we can predict for example the velocity of a star from the shifting of the emission spectrum of its atoms (see section 3.2). There is however no continuous oscillation of period λ attached to a photon and there is no modification of its velocity. The speed and the way a photon evolves are the same in all inertial frames, but, experimentally, we register a different perception of the same photon if we move with respect to it. With the usual Fourier analysis, the explanation of this behavior is unclear, since the photon turns out to consist of a single pulse and there is no frequency directly associated with it. In fact, what changes is the longitudinal spreading of the signal carried and not the modality of transportation.

A picture taken at a given time of a specific photon emitted by a known atom in the passage from a certain state to another, can tell us if the atom is moving and what its velocity is. We get this piece of information even considering that a picture offers a stationary view. The dynamics of the photon is unessential, being a free-wave it is enough to recover the initial orientation of the electromagnetic fields to determine its future evolution (it will shift at the speed of light in the direction of the cross product $\mathbf{E} \times \mathbf{B}$). Such an observation attracts the attention to the geometrical properties of the photon (see section 1.6). The mechanism of the *Lorentz contraction* of lengths, of not immediate explication in our everyday life, assumes a simplified form when dealing with the different representations of the same photon as a function of the observer. In fact, instead of comparing what happens at velocities $\mathbf{V}$ and $\mathbf{V} + \mathbf{v}$, where $|\mathbf{V}| = c$, one just sits on the photon and looks at its variation with respect to the parameter $\mathbf{v}$. Better examining a steady picture than trying to catch a photon with unconvincing meters and clocks. To the reader who remarks that the picture of a photon cannot be taken in reality, I answer that a mathematical shot is available from the analysis of the explicit solutions of my model equations (see (83) in appendix E).

In conclusion, a given photon that always moves at the speed of light may appear to us to be elongated or contracted depending on the velocity of the source. Sources escaping from us emit photons that, when reaching our instruments, display a lower energy (*redshifting*). Light emanated from galaxies is usually affected by redshift. Cosmologists tend to attribute the explanation of this phenomenon to the expansion properties of the universe. Less accepted is the idea that such a modification could be also a consequence of some "accident" that occurred during the journey, so that an anomalous shape of the photons might not be directly associated to the movement of the emitter[27],[28]. If this last was the case, what do we really see when we observe the universe?[29]. I will add further personal considerations in section 3.8.

Finally, going back to the study of invariance properties, the result that a free photon remains a free-wave for all the inertial observers is a wonderful feature, emphasizing the absolute nature of such an entity. The property is analogous to that stated by Einstein (see footnote 26) concerning the Lorentz invariance of the notion of a non-accelerated charge. Let me remark anyway that a photon is something rather different from an electric charge. First of all, it does not radiate, but the electromagnetic signal remains inside its traveling domain, without diffusing. In agreement with this, though $\rho = \text{div}\mathbf{E}$ is usually different from zero in the photon's support, the integral of ρ vanishes (because of the zero boundary conditions on $\mathbf{E}$), respecting in this fashion the *Gauss theorem*. As I will explain in the coming sections, photons may possibly transmute into real charges. This potentiality becomes effective as one tries to confine them in a bounded

---

[27] From [Misner, Thorne and Wheeler(1973)], p.775: "No one has ever put forward a satisfactory explanation for the cosmological redshift other than the expansion of the universe. The idea has been proposed at various times by various authors that some new process is at work ('tired light') in which photons interact with atoms or electrons on their way from source to receptor, and thereby lose bits and pieces of their energy. Ya. B. Zel'dovich gives a penetrating analysis of the difficulties with any such ideas: ...".

[28] From [Awada, Alia and Majumderd(2013)]: "According to the theory of the *rainbow universe*, photons travel on slightly different paths through spacetime, depending to their energy. The theory does not necessitate of a singularity point in time when the universe is infinitely dense, disclosing the possibility that a Big Bang may never have happened".

[29] From [Ijjas, Steinhardt and Loeb(2017)]: "Yet even now the cosmology community has not taken a cold, honest look at the big bang inflationary theory or paid significant attention to critics who question whether inflation happened. Rather cosmologists appear to accept at face value the proponent's assertion that we must believe the inflationary theory because it offers the only simple explanation of the observed features of the universe".

region of space. The study of such a situation is going to be extremely complicated and the instruments of special relativity are not fine enough for a deeper analysis. This is mainly due to the fact that my photons have an internal structure. For example, later, I will assume that parts of a photon may travel at different speeds, so breaking the second postulate of special relativity. The explanations will rely on the possibility of handling non-inertial reference frames. The spontaneous battleground to look for an upgraded theoretical validation of the model is *general relativity*, where the notion of invariance gets more sophisticated. This will open the path to the study of constrained waves and to my philosophical understanding of nature's facts. Consequentially, the concepts of space-time curvature and gravitation will emerge.

1.5   The geometry of space-time

Thanks to the work of distinguished mathematicians, such as B. Riemann and G. Ricci Curbastro, Einstein was able to further extend the theory of relativity by expressing physics laws in tensor form. I guess that, initially, the idea was to study more complicated changes of variable in the space-time, such as (non-inertial) uniform accelerated frames, but the developments turned out to be impressively broad, consistent and elegant. Personally, I consider the theory of *general relativity* to be the best achievement of mathematical physics. Although the scientific literature is rich of variants and extensions, I believe that the original Einstein's version is already complete and sufficiently concise. Moreover, there are still many unexplored territories, to which I would like here to concentrate my attention.

It is not easy to describe in a few non-technical paragraphs the principal ideas, taking also into consideration that I am far from being an expert; nevertheless I will try here to come out with some naive explanations. Such an introduction is not probably the one usually followed in classical treatises, however it is coherent with my approach to the study of electromagnetism. For these reasons and for the sake of brevity, I will omit several canonical aspects, that can however be retrieved from specialized publications.

To start with we may examine for instance the case of a uniform accelerated frame along a straight path. Physics laws from this viewpoint are going to be formally different from those related to an observer considered "at rest". By the way, a general common formulation can be provided. This is obtained by first noting that the two observers can be supposed to live in different "space-time geometrical environments". Here the meaning of the term geometry is rather complex. Let me be vague for a while. With some adjusted measuring instruments an observer locally regulates the detection and the interpretation of his own results. The metric system subtly depends on the environment and may not coincide with that of another observer. For this reason, the two of them may have a different perception of the facts going on. Nevertheless, the governing laws are written in some unified abstract form, that takes into account the presence, well concealed, of possible geometric backgrounds. In this way, the laws maintain an absolute validity, however they can assume various appearances in the presence of different observers. Let us make these rough intuitions more clear with a standard example.

Typically, one begins with analyzing what happens on a free-falling elevator, subject to a gravitational field, as far as an external observer at rest is concerned. Another observer, placed in the elevator, may see free-falling objects at rest. By conducting simple experiments of dynamics on these objects, there is no way the falling observer can realize he is in accelerated motion. According to him the rules of mechanics are the usual ones. We can then associate with the elevator an isotropic space having a "flat" geometry, which is basically the standard Euclidean space we are used to. This is true at least in the elevator's micro-universe, i.e., the flat properties of the space-time hold up to the elevator's boundary. Also for the external observer the rules of mechanics are the same, except that the contribution of a suitable constant vector (proportional to a given acceleration) has to be added to the formulas. The novelty of the Einstein's approach is to include this gravity-like vector in the geometrical environment of the second observer. The presence of this term is going to provide his space-time with a "curvature". According to Einstein's idea, such a deformation of the geometry has to be attributed to the presence of some far-away large mass, which is exactly the one causing the elevator to fall. In the end, in both cases (the flat and the curved ones), Newtonian laws can be applied (at least, locally) with no additional corrections, because these are implicitly and automatically assigned to the geometric parts.

Let me go a bit further into the details. The space-time geometry is described by the so called *metric tensor*. Considering time as the first of four dimensions, for a given system of coordinates, the metric tensor is expressed by a 4×4 symmetric matrix $g_{\alpha\beta}$ containing distinctive point-wise information about lengths, angles and the way time develops. Here, the 4D background is a natural choice. Indeed, already in the theory of special relativity, there is no absolute notion of space and time as individual entities, because these turn out to be fully entangled. In section 1.4, my excursion on special relativity did not cover this issue much, that can be however sharpened by searching in the appropriate texts. In the current exposition, the link between space and time starts assuming more importance.

Based on the example discussed above, two different metric tensors can be assigned to the observers (the one at rest and the one in the elevator). The next step is to find the way to express the laws of mechanics in terms of a generic metric tensor. The final step is to check whether there is a closed form to write the ruling equations, by hiding their dependence on the metric tensors, obtaining a universal formula. In this fashion the laws turn out to be written in the so called *covariant form*. When necessary, specific practical versions can be recovered from such a symbolic system, by plugging the proper metric tensor in the general formula and reading the laws in the local geometrical space. If the space is flat, the corresponding metric tensors are uninfluential; thus one gets plain Newtonian laws (presenting no additional forcing terms), describing what happens in the elevator. This is actually true with good approximation in a range of velocities much smaller compared to the speed of light. When the space is modified by the gravitational context, one gets the laws of the observer at rest, differing from the previous ones by the addition of an acceleration field. To achieve mathematically this goal, it is necessary to create a differential calculus that clarifies how to compute the derivatives in an abstract way. Superficially, such *covariant derivatives* appear with an innocent symbol of differentiation. When effectively evaluated, they make heavy use of the entries of the metric and the system of coordinates defined in the specific geometric environment[30].

There are infinite metric tensors that can be associated with a given geometrical space exhibiting a certain property (flatness for instance), depending of all possible ways to define systems of coordinates on it. Therefore, a flat space remains flat even by defining on it "curvilinear" coordinates. A change of variables is a transformation of the 4-dimensional space into itself, and depends on four functions, while the metric tensor (which is symmetric) can accommodate up to ten distinct functions. Therefore, there are infinitely many typologies of geometrical spaces that cannot be

---

[30] From [Misner, Thorne and Wheeler(1973)], p.387: "The transition in formalism from flat spacetime to curved spacetime is a trivial process [...]. But it is nontrivial in its implications. It meshes gravity with all the laws of physics. Gravity enters in an essential way through the covariant derivative of curved spacetime".

globally transformed one into the other by a mere change of coordinates. Under these circumstances, one can start distinguishing among all possible "curved spaces". The presence of masses and the origin of gravity are considered to be the most relevant motivations yielding curvature. The Einsteinian *principle of equivalence* actually states that a non-inertial reference system is equivalent to a certain gravitational field. On the other hand, one may have plenty of intermediate situations where the curvature of the space does not clearly descend from the existence of well-determined masses. I am mainly concerned with these cases and I will continue to use the word "gravitation", even if sometimes the term is not the proper one.

Technically, the concept of curved space applies at a global level. It is possible to adapt locally the system of coordinates in order to have the impression of being for instance in a flat space, but the procedure might not be extendable to the whole manifold[31]. It is then far from being easy to decide about the properties of a space associated with a given metric tensor. The most straightforward method recalls the one used in basic calculus, that amounts to evaluating second derivatives. In the case of 4×4 tensors this is quite a tricky and cumbersome way to proceed, but operatively is the only reliable one. Anyway, it is not my intention to proceed with annoying notations and computations, so I will continue to keep the exposition as simple as possible.

Before going ahead, it is important (and necessary) to embed my theory of electromagnetism in the context of general relativity. In other words, it is compulsory to state my model equations in covariant form. Nevertheless, I do not want to here make use of tedious technicalities. Let me try anyway to provide some quick explanations. The shortest way to write down Maxwell's equations with the help of 4×4 tensors is: $\partial_\alpha F^{\alpha\beta} = 0$, for β ranging from 0 to 3 (see (13) in appendix A). In a more cryptic form one can write: $\partial F = 0$. Here F is the *electromagnetic tensor* and ∂ is a suitable covariant differential operator. Both of them depend on a given metric tensor $g_{\alpha\beta}$, although this is not visible from the compacted formula. For example, when $g_{\alpha\beta} = \mathrm{diag}\{1, -1, -1, -1\}$ is the flat metric tensor in Cartesian coordinates, the expression $\partial F = 0$ becomes (12) in appendix A. Readers with a background in calculus can easily go over the computations and check that this writing

---

[31] From [Misner Thorne and Wheeler(1973)], p.191: "Geographers have similar problems when mapping the surface of the earth. Over small areas, a township or a county, it is easy to use standard rectangular coordinate system. However, when two fairly large regions are mapped, each with one coordinate axis pointing north, then one finds that the edges of the map overlap each other best if placed at slight angle".

really leads to Maxwell's equations. More precisely, a couple of Maxwell's equations ((1) and (3)) directly descend from this approach, while the two other equations ((2) and (4)) are implicitly assumed when constructing F. Now, for a certain choice of the metric tensor, a full rendering of formula $\partial F = 0$ may look weird and unusual, but still it represents the description of a given electromagnetic phenomenon from the viewpoint of the observer embedded in that specific metric space.

As far as my equations are concerned, their contracted form looks as follows: $\partial F = (\rho/c)V$ and $FV = 0$ (see (25) and (26)), where V is a 4-component velocity vector. The first equation is the Maxwell's version with a current term, though we know from section 1.3 that such a term is not due to external sources, but the current is actually part of the modeling of a wave. The second equation expresses the free-wave condition. Both writings are well-suited to the framework of covariant calculus. As already stressed in previous sections, my formulation does not add any new technical difficulty and can be naturally inserted in existing theories. The relevant advantage is to be able to work with an extended space of solutions, preserving the features of classical electromagnetism.

According to appendix D, another method to write down Maxwell's equations in relativistic fashion is to use the electromagnetic 4×4 *stress tensor* U (see (33)), providing information on energy fluxes and momenta in electrodynamics processes. Conservation of these physical quantities is imposed by setting to zero the covariant derivative of the stress tensor, i.e.: $\partial U = 0$ (see (34)). This symbolic expression turns out to be the essence of all energy-preserving electromagnetic phenomena, cleared from specific geometric environments and coordinate systems. One has the following implication: $\partial F = 0 \Rightarrow \partial U = 0$ (see (37)), saying that Maxwellian waves comply with energy conservation rules. The good thing is that Maxwell's equations are not the sole laws compatible with $\partial U = 0$, though (unlikely) the interest is reduced to such a linear subcase in common practice[32]. The computed quantity ∂U is the sum of several terms; it is zero if Maxwell's equations hold true, but it is also zero when plugging my set of equations (and they are probably the more general group of equations able to do this).

---

[32] From [Donev and Tashkova(2012)]: "... the vector $c^{-1}\mathbf{E} \times \mathbf{B}$ [...] appears as natural complement of Maxwell stress tensor, and allows to write down dynamical field equations having direct local energy-momentum balance sense. However, looking back in time, we see that this viewpoint for writing down field equations has been neglected, theorists have paid more respect and attention to the linear part of Maxwell theory, enjoying, for example, the *exact* but not realistic, and even *physically senseless* in many respects, plane wave solutions in the pure field case".

The verification is simple; it is enough to keep the quantity ρ different from zero and perform some algebraic manipulation. More insight is given in appendix D and [Funaro(2009b)]. This little proof is another crucial point in favor of my approach. In perfect line with existing and well-assimilated material, I am able to provide an alternative way of organizing facts. Some readers may remark that I am just putting together trivial things, already well-established. Indeed, my work does not seek to be in conflict with consolidated achievements, but at the same time photons are now fully entitled to enter the world of classical electromagnetic phenomena.

Let us now come back to the elevator problem. Each observer is locally embedded in his geometrical environment, and, effectively, for confined regions and small velocities, the mechanical behavior of bodies may be described with good approximation by Newtonian laws, so that, to a first level of approximation, the interpretations of nature's facts by the various independent observers are in agreement. For small velocities, the observer at rest can "see" what happens in the elevator, using for instance light as a means to capture information. This becomes less true as velocities increase. A more serious analysis is then unavoidable, since we are in the process of studying photons more closely.

From these thoughts we learn that the geometry of the frame at rest and that relative to the elevator might coexist in the same universe, but a smooth passage is difficult to imagine without crossing some kind of boundary (placed on the elevator's cabin walls, for example). The problem is usually resolved by assuming the space-time to be fully covered by overlapping *coordinate charts*, but what really happens in the common regions is a bit unfocused from the physics viewpoint[33]. For instance, outside the elevator's cabin the surroundings are affected by gravitation presence, while inside the effect is counterbalanced by the motion of the elevator. But, how is the space-time geometry in the immediate external neighborhood of the cabin? What changes are involved if I jump from the falling elevator to the stationary ground? Usually, the stop is quite abrupt (don't try this at home!). Does this mean that in such a process I have been crossing a surface of discontinuity? Actually, something discontinuous happened: I

---

[33]From [Bergmann(1968)]: "A free-falling frame of reference differs, however, from all frames of references previously discussed in that it cannot be extended arbitrarily far through space and time. It is well defined only in the neighborhood of a world point, in a limited region of space and for a limited period of time. Because of the inhomogeneity of all gravitational fields, any attempt to extend a free-falling frame of reference to far distances or to long periods of time must fail through internal inconsistencies".

gained *weight*, a feature that was not present when in the elevator. These foundational issues are often disregarded in the general relativity literature, in favor of more practical aspects. They have to be treated with caution, since it is easy to be caught in a trap. I also do not have sufficient experience to handle them, but I understand they hide something interesting and worth developing. I will have a chance to return to such important questions in due course (see sections 2.2 and 2.6).

Afterwards, one is left with the nontrivial subject of deciding what are the properties of the entire 3D universe (4D including time)[34]. The question is pertinent but the answer is not that easy. The problem of finding the constitutive geometry of our universe is a troubled topic that involves age-old debates. Large astronomical masses produce a deformation of the 4D space-time (from its hypothetical uniform flatness), having decisive effects on its metric properties, but nobody at the moment knows what is the most plausible occurrence. Cosmological models, including or not *Big Bang theories*, proliferate. It is not my intention to engage in the contest, but I will advance an explanation of the facts, that is very far from the motivations behind these struggles. The path is still long, and it will take time to prepare the ground. Let me say for the moment that, due to the reasons put forward in the previous paragraph, I believe there is no way to recover a global smooth description of the geometrical properties of the whole universe, and this is not only due to the difficulty of finding suitable asymptotic expressions at some far-away boundary[35]. My conjectures go instead in the opposite direction, where the universe may be fragmented into a myriad of individual geometrical settings, akin to lower-level Leibniz monads, separated by appropriate boundaries[36]. This should not prevent

---

[34]From the introduction in [Fock(1959)]: "In order to construct a theory of gravitation or to apply it to physical problems it is, however, insufficient to study the space and time only locally, i.e. in infinitely small regions of space and periods of time. One way or another one must also characterize the properties of space as a whole. If one does not do this, it is quite impossible to state any problem uniquely. This is particularly clear in view of the fact the equations of the gravitational, or any other field, are partial differential equations, the solution of which are unique only when initial, boundary or other equivalent conditions are given. The field equations and the boundary conditions are inextricably connected and the latter can in no way be considered less important than the former".

[35]From [Fock(1959)], p.396: "... In the first place there is an incorrect initial assumption. Einstein speaks of arbitrary gravitational fields extending as far as one pleases and not limited by boundary conditions. Such fields cannot exist".

[36] From [Leibniz(1714)], n.7: "Further, there is no way of explaining how a Monad can be altered in quality or internally changed by any other created thing; since it is impossible to change the place of anything in it or to conceive in it any internal motion which could

the search for an all-comprehensive model, bypassing the micro-structures by using the right amount of approximation.

At the moment, a more basic question can be raised: what is the cause responsible for a certain geometrical setting? In the case of the falling elevator, one has to assume the existence of a given gravitational field, due to the presence of some distant mass. Therefore, at least to a first approximation, a neat connection can be established between gravitation and the deformation of the space-time. According to Einstein, a system of masses produces a distortion of the geometrical background. Within this context, other point-wise masses may move along geodesic paths and, if their velocities are relatively small, they do it in agreement with Newtonian mechanics (as planets do around the Sun). But Einstein's results run much deeper. He was able to write down a complicated nonlinear tensor equation, providing the appropriate metric tensor for a given source of energy. Such a source can be either represented by a set of masses or an electromagnetic type phenomenon (in the last case we need to recall the stress tensor U mentioned a few paragraphs above).

Einstein's equation (see (40) in appendix D) involves second order derivatives of the unknown metric tensor, recalling in this way the concept of curvature. It is the mother of all deterministic physics laws, a concentration of wisdom and elegance. I am not surprised that, after such an amazing achievement, Einstein felt himself close to the formulation of a comprehensive mathematical model for all nature's events. Regrettably, there were still two big open questions to solve; the first one was the non-trivial goal of giving an explanation to quantum mechanics, the second one, of which he was not aware, was the inconsistency of Maxwell's model. We will see that fixing the second issue brings clarification of the first.

Let me spend a few more words on Einstein's equation. After representing space-time with the help of a suitable metric tensor $g_{\alpha\beta}$, an almost exhaustive estimate of the curvature is obtained by computing the *Ricci tensor* $R_{\gamma\delta}$, based on second order differentiation. This is then used to build, through slight modification, another 4×4 tensor $G_{\gamma\delta}$. If for instance $g_{\alpha\beta}$ represents a flat space, then $G_{\gamma\delta}$ is identically zero. Suppose that we

---

be produced, directed, increased or diminished therein, although all this is possible in the case of compounds, in which there are changes among the parts. The Monads have no windows, through which anything could come in or go out. Accidents cannot separate themselves from substances nor go about outside of them, as the 'sensible species' of the Scholastics used to do. Thus neither substance nor accident can come into a Monad from outside".

want to determine the effects of a given forcing source tensor $T_{\gamma\delta}$ on the geometric environment. Now $g_{\alpha\beta}$ is the unknown and Einstein's equation amounts to solving a complex nonlinear system of equations recoverable by the relation: $G_{\gamma\delta} = -\chi T_{\gamma\delta}$, where χ is a given constant. Both the equated terms turn out to be implicitly dependent on $g_{\alpha\beta}$. In absence of sources, the space turns out to be flat (more precisely, with zero Ricci curvature). Alternatively, the presence of energy (in principle of any kind) is able to ply the geometric background and provide it with a nontrivial curvature.

The dynamics of small inoffensive slow-moving objects is affected by a modified geometric environment and their evolution is ruled by universal laws. Of course, this is a simplistic interpretation, since any object carries energy in some way and this further contributes to the space-time deformation and to the successive evolution of itself and other entities. But, the above is just a technical difficulty; as a matter of fact, Einstein's equation is general enough to handle any type of situation. As far as masses are concerned, under the hypothesis that their development is within certain limits, Einstein's equation is able to reproduce the laws of gravitation, and this is why it has a primary role in cosmology, provided certain simplifications are made[37]. However, my interest here is predominantly focused on electromagnetic applications, while I will return to the question of masses in the next chapter. It is my opinion that the potentialities of Einstein's equation have not been fully exploited and it is my intention to analyze new directions of development. Unfortunately, since I must deal with luminous phenomena, I cannot count on certain assumptions that have been considered valid in the cosmological analysis.

1.6 The geometry of a single photon

Going back to the elevator example considered in section 1.5, one can also examine the trajectory of a photon from the two different viewpoints. Let us suppose that the photon, emitted by a tool traveling with the elevator, is shot horizontally (i.e., perpendicular to the direction of fall). As far

---

[37]In [Einstein(1920)], the author writes: "According to the general theory of relativity, the geometrical properties of space are not independent, but they are determined by matter. Thus we can draw conclusions about the geometrical structure of the universe only if we base our considerations on the state of the matter as being something that is known. We know from experience that, for a suitably chosen co-ordinate system, the velocities of the stars are small as compared with the velocity of transmission of light. We can thus as a rough approximation arrive at a conclusion as to the nature of the universe as a whole, if we treat the matter as being at rest".

as the observer at rest is concerned the photon will not follow a straight line: indeed, it is "falling down" together with the elevator (the effect is almost imperceptible, considered that light flashes at a speed approximately equal to 3×10^8 meters/second). This somehow entails that the path of an electromagnetic wave is affected by a curved geometry due to the exterior gravitational field.

Simultaneously, the photon itself, with much less emphasis, modifies the surrounding geometric environment. The fact that photons can be deviated by the presence of masses (*light bending*) was confirmed by a decisive experiment a long time ago, providing full recognition to the theory of general relativity. As a mathematician I consider the work of Einstein to be excellent also without this confirmation, but, of course, the verification of certain predictions is part of the game in physics. In some ways we can see the photon as a carrier of energy and, by formally assigning a mass to it (although a photon is considered to be massless), according to the famous relation $E = mc^2$, we could roughly interpret the change of trajectory as an interaction of masses.

Nevertheless, I think this version of the facts may be reconsidered remaining within the limits of general relativity. Einstein's equation, being the fulcrum around which electromagnetism and gravitation rotate, can actually offer a more precise description of the real behavior. To this end, it is also worthwhile recalling that the photon is not just a point; it has an internal structure and, at the same time, is an electromagnetic wave described by my model equations. Moreover, according to the heretical viewpoint that I will try to defend in the sequel of this exposition, a gravitational field only exists as a by-product of suitable electromagnetic phenomena, so that, in a general context, its interaction with a photon is exquisitely of electromagnetic type. To this we will have to add some quantum effects, that make the story more intricate. These audacious statements may sound rather speculative at the moment, but I will do my best to clarify the situation later. For the moment, the first goal is to investigate how the passage of a free-wave can alter the stillness of a flat universe.

Recall that a photon is a free-wave which is individuated by the shifting at the speed of light of a certain distribution of electric and magnetic fields (see figure 1.2). The motion is precisely described by my set of model equations (see the exact solutions of appendix E). As I said in section 1.5, it is possible to assign to this phenomenon a 4×4 stress-energy tensor U, whose entries are quadratically dependent on $\mathbf{E}$ and $\mathbf{B}$. Such a tensor can be written, in covariant form, as a function of a generic metric tensor. In a successive step, for a given constant μ, one can plug the source term $T = -(\mu^2/c^4)U$ (see (56)) on the right-hand side of Einstein's equation and try to solve the corresponding differential system in order to find the metric tensor. Recall that the left-hand side contains the Ricci tensor, i.e., a concomitant of the second derivatives of the metric tensor to be found. If things are done properly, one can come out with a solution, suitably represented by a tensor $g_{\alpha\beta}$ (note that there is no unique solution to Einstein's equation, although families of solutions can be considered qualitatively equivalent, if they just differ by coordinate changes). In the most complicated situation there are 10 scalar unknowns, that are functions of both time and the three space variables. These unknowns are all coupled by 10 nonlinear scalar equations. In realistic situations, the problem can only be approached by numerical simulations, but still its complexity may be out of reach for many computational codes. There are a few known explicit solutions mostly related to concentrated masses, and, except for some theoretical analysis, the system of equations remains a morsel hard to chew.

It has to be remarked that the interest in solving Einstein's equation with electromagnetic sources is very limited. Real-life applications are mainly astronomical, where electromagnetism interferes with negligible effects. There are coupled models where, for instance, planetary systems are studied in conjunction with magnetic phenomena, but they are far from pure electromagnetism. Solving Einstein's equation with an exclusively electromagnetic forcing term is primarily of interest to the academic world and there are plenty of results, although they do not strike as deep as those I am going to examine. The discussion that will follow is aimed at stimulating attention towards relativistic electromagnetism and is crucial for understanding basic questions that are parts of the backbone of my general construction.

The possibility of getting explicit expressions of the metric tensor in terms of $\mathbf{E}$ and $\mathbf{B}$, in the case of free-waves, is an achievement obtained in [Funaro(2008)] and generalized in [Funaro(2009b)] (see also appendix E). Therefore, with the help of that analysis, I can clearly understand what is going on in and around a photon. To get the result I used symbolic manipulation programming and considered that the orthogonality of $\mathbf{E}$ and $\mathbf{B}$ may, using an appropriate coordinate choice, involve only two components different from zero. In the end, recalling that the intensity of $\mathbf{E}$ is c times that of $\mathbf{B}$, we are left with only one degree of freedom. For the sake of simplicity, it is also possible to suppose that $\mathbf{V} = (0, 0, c)$, expressing the fact that the wave develops along the direction of the third space coordinate, that can be denoted by z (as for instance in the case (49) treated in appendix E). Let me skip here the technical problem of converting these ingredients in the language of four-components vectors. Afterwards, I have been searching for a very plain version of the metric tensor $g_{\alpha\beta}$ (basically a diagonal matrix). Roughly speaking, it turns out that on each wave-front the "intensity' of the metric deformation, along the transverse direction to the propagation vector $\mathbf{V}$, is directly related to the magnitude of $\mathbf{E}$ (proportional to that of $\mathbf{B}$). Along the longitudinal shifting direction, the entire solution of Einstein's equation for free-waves can be reduced to a single ordinary differential equation of the form: $-\sigma''\sigma = g^2$ in the variable $\xi = ct - z$, where t is the time and g is a given function related to the profile of the wave along the z axis (see (61); here the constant $\chi\mu^2/c^4$ has been omitted for simplicity).

For the success of this computation, there are two other important aspects that have to be taken into consideration. First of all, the stress tensor depends on the unknown geometry, so that the metric tensor appears both on the left-hand side (via a complex nonlinear system with second-order partial derivatives) and on the right-hand side to accompany the source term. I say this because the dependency of the right-hand side on $g_{\alpha\beta}$ is systematically neglected, although essential to recover a simple and elegant final answer. The second aspect concerns the choice of the sign of the source term, since a switch considerably changes the nature of the analytic problem. As a matter of fact, with the proper choice, one can get significant solutions from: $-\sigma''\sigma = g^2$. For instance, if $g(\xi) = \sin\omega\xi$, a possible solution is $\sigma(\xi) = \sin\omega\xi/\omega$. Assuming instead the "wrong" sign for the right-hand side tensor, one arrives at the equation $\sigma''\sigma = g^2$ that has no meaningful solutions (most of them being unbounded or not compatible with homogeneous boundary conditions). The last was anyway the version analyzed in the past and solutions were found with an abundant misuse of ramshackle adjustments and approximations (see, e.g: [Bondi, Pirani and Robinson(1959)]). In [Funaro(2008)], p.92, explanations are provided in favor of the first choice, and further hints will be given in this chapter. Let me point out that this alteration does not contradict any physical principle.

Let us go now to the real reasons for this investigation, i.e., understanding the anatomy of a photon. I will discuss the qualitative consequences emerging from a rigorous study, that I am not reporting here. In vacuum, far away from perturbing fields of any kind, the natural metric environment is associated to a total flat geometry (immutable meters and invariable clocks everywhere at every moment; zero curvature tensors). In this environment the electrodynamic modeling equations coincide with those I have introduced. In this landscape, during a photon's passage, together with the electromagnetic wave one can also detect a traveling *gravitational wave*. Such a secondary wave has the same support as the original electromagnetic one, hence it travels in a straight line at the speed of light. Note that, in order to preserve the global continuity, **E** and **B** (transverse to the direction of propagation) tend to zero at the border of the support of the photon, outside of which the signal vanishes. The same is done by the gravitational signal. If in the longitudinal elongation the photon intensity is modulated by a piece of sinusoid, the longitudinal intensity of the gravitational perturbation also behaves as a sinusoid; in fact, they can be linked via the above mentioned ordinary differential equation. When I use the word "gravitational" I mean it in the framework of alterations of the metric in the context of general relativity. There are no masses necessarily involved. The presence of the gravitational wave tells us that there is a local momentary deformation of the space-time at some points of the flat universe, situated in the path of the wave. Such a wave has no influence outside the photon's boundary.

Now, we can enter the photon. Here, we find a different metric, i.e., the one generated by the electromagnetic wave through Einstein's equation. One may start wondering what the constitutive model equations look like (recall that they come from a general covariant writing, but they can be specialized depending on the current geometrical setting). The result is quite surprising: an observer placed inside the wave does not realize that he is in motion and the local metric tensor deforms the electromagnetic field in such a way that it looks constant in all transverse directions. The ruling equations are of the form $0 = 0$, hence they are totally useless. Basically, the traveler is convinced that he is on a plane wave of infinite extent. This is a consequence of the correction of the metric, but the inner observer is not aware of this. Correspondingly, both the divergences of **E** and **B** turn out to be zero, so that one finds himself in a stationary Maxwellian world. Therefore, if for an external observer a photon is a confined object of finite measure, the internal observer thinks he is in an isotropic space of infinite extent. As a matter of fact, as the internal observer approaches the boundary, the metric correction modifies his perception of distances, and the border remains a far horizon[38]. Jumping from a running photon (if

---

[38]From *Alice's Adventures in Wonderland*, by L. Carrol: "Well, in our country — said Alice, still panting a little — you'd generally get to somewhere else, if you ran very fast

this were possible) to the total calm of an otherwise unperturbed universe is quite a shock (like jumping from the elevator of the previous section). There is no smooth passage, but one has to overcome a kind of spatial singularity, dividing two independent universes.

Riding a photon was probably the dream of Einstein. Simulating this journey was one of the aims of the theory of relativity. According to my results, a photon is not a point, but has a size and some interior architecture. Under this assumption, thanks to Einstein's theory, I am able to get accurate descriptions of all the fields involved. The amazing conclusion is that each photon is an isolated flat unbounded universe with no internal relevant organization. Observers living in it cannot conceive of an external world. This indivisible entity can be effectively called "elementary particle" and is a real "atom" in the Democritus sense (see also footnote 36). Coincidentally, it is also a wave. From a philosophical viewpoint we should not concern ourselves anymore with the photon's internal structure; in the end such an entity is only made by diaphanous vector fields, not necessitating any subordinate structure.

A deeper study shows that, from outside, the shape of a photon is isomorphic to a toroid[39] that moves in the direction of its axis (see figure 1.2). This is mainly due to the div$\mathbf{B} = 0$ condition. By the orthogonality of $\mathbf{E}$ and $\mathbf{B}$, both fields must vanish at some point inside each wave-front. This internal region of null vectors can be assumed to be as large as one pleases. As we will examine later, rings are the bricks of our universe. Let me finally note that the metric deformation automatically adapts itself to the electromagnetic signal. This occurs almost independently of the field organization inside the photon. In this way, we can attribute to the particle topological properties that do not follow from the signal carried, while its internal structure invariably remains of some uniform "density". In the end, the only peculiar and observable property of a photon will be its energy.

As we saw the electromagnetic fields of a free-wave (including photons) are not alone in their trip, but they share their space with a gravitational

---

for a long time as we've been doing [...] A slow sort of country! Now, here, you see, it takes all the running you can do, to keep in the same place. If you want to get somewhere else, you must run at least twice as fast as that!".

[39]From [Thomson(1935)]: "It occurred to me later that a great many difficulties could be got over if we suppose that the photon, instead of being a single ring of electric force, consisted of a harmonic train of such rings; such a train would have a wave-length of its own, and could produce interference by itself without the assistance of waves of another kind". These conclusions by J.J. Thomson are still technically far from the final answer. The intuitions were however focused in the right direction.

type wave, i.e., a shifting local alteration of the metric properties of the flat space hosting the wave. One may ask if the gravitational effects may exert a mechanic influence on nearby objects. My personal opinion is definitely affirmative, but the intensity of the perturbation may strongly vary with circumstances. We will have time later to discuss the so called *radiation pressure*, a phenomenon that characterizes the interaction between light and matter. In order to do this, it is necessary to generalize the model equations, including a pressure term as introduced in fluid dynamics. In the *Nichols radiometer*, a device experimented with at the beginning of the *XX* century, focused light induces measurable torsion on a wire attached to a silvered plate. Despite its simplicity, this phenomenon is very difficult to explain in a classical context of deterministic differential equations, since it actually involves the transmutation from electromagnetic to mechanic forces. Before facing such an intriguing issue, let me draw here some basic conclusions, that may lead to more profound reflections.

By solving Einstein's equation in the case of sinusoidal modulated $(g(\xi) = \sin \omega \xi)$ wave-fronts in the direction of propagation, we discovered that the corresponding gravitational wave is also of sinusoid type $(\sigma(\xi) = \sin \omega \xi / \omega)$, but the most important thing to remark is that it displays an intensity that is inversely proportional to the sinus frequency (see also (62) in appendix E). Thus, high-frequency electromagnetic waves produce very low-intensity gravitational waves. Tiny photons, as those for instance associated with nuclear or atomic phenomena, can be related to short wave-lengths and, therefore, high frequencies. Their gravitational effect on bigger objects is certainly negligible. The situation is going to be different for very slow time-varying electromagnetic phenomena. In this case I am thinking about wave-lengths at the level of astronomic distances. We know that this is the kingdom of real-life gravitation. After stating this, I do not want right now to conclude that gravitational effects are a by-product of low-frequency electromagnetic waves. We actually need more material in order to arrive at that point, but this is a first clue in favor of such an interpretation.

From the practical point of view, the detection of low-frequency gravitational waves accompanying an electromagnetic wave is hard to accomplish. Since they travel at the speed of light, the size of the measuring instruments must be outrageously huge in order to catch an appreciable signal. Producing low-frequency electromagnetic waves is also a troubled problem, due to the exaggerated dimension of the emitting devices. High frequencies, as I said, should not produce relevant gravitational modifications. Perhaps, some experiment could be conducted by using relatively small transmitting devices, based on a high-frequency carrier, to transport a lower frequency message. The existence of gravitational waves, not directly associated to electromagnetic phenomena, is predicted by Einstein's equation after suitable linearization. Their practical detection is a recent achievement. The subject is complex and has been widely investigated. In my approach, gravitational waves may survive with some independence, but always within the context of an electromagnetic background and principally relying on the nonlinear features of all the equations involved (see [Aldrovandi, Pereira and Vu(2007)]). Therefore, once again, I will stay away from the standard path.

Let me finally remark another interesting fact. Each point in the wave is associated to a triplet $(\mathbf{E}, \mathbf{B}, \mathbf{V})$. For a free-wave such a system of vectors in the 3D space turns out to be orthogonal and may represent a local reference frame, shifting in time with the wave. It is nice to observe that the pointwise metric tensor, obtained by solving Einstein's equation for the given wave, is somehow compatible, in its simplicity, with the triplet $(\mathbf{E}, \mathbf{B}, \mathbf{V})$. An electromagnetic signal is then in relation with a multitude of point-wise frames, that simultaneously define a system of coordinates and a metric (the space unity of measure is proportional to the intensity of the transverse electromagnetic field). The time variable does not cause problems due to the fact that there is only one velocity involved. Hence, there exists a kind of isomorphism between the geometric environment naturally associated to the wave and that recovered via Einstein's equation.

In vacuum, with a total absence of signals, there is no way to understand the geometric property of the space-time, that, for this reason, can be considered flat. As a photon passes by, the exclusive presence of the electromagnetic fields guarantees the existence of a modified geometry, both in the form of referring triplets $(\mathbf{E}, \mathbf{B}, \mathbf{V})$ and in the Einsteinian sense (the two things seem to be formally indistinguishable, at least in the case of free-waves). These observations, waiting for a more formal theoretical analysis, suggest the following conjecture: vector fields accompanying a wave are not only the source of a geometric environment, they <u>are</u> the geometric environment! This should also imply that Einstein's equation provides for a surplus of information, the sole knowledge of the electromagnetic fields being sufficient to understand the gravitational scaffolding. I will look for further confirmations of this hypothesis in the next section. Let me anticipate that in order to get a result of this kind, in addition to the triplet $(\mathbf{E}, \mathbf{B}, \mathbf{V})$, one needs the help of a scalar potential p (which turns out to be zero in the case of free-waves), corresponding to the previously mentioned "pressure". This will actually constitute the missing link between electromagnetism and mechanics.

Later, I will claim that in our universe there is nothing but electromagnetic phenomena. In fact, gravitation, interpreted as a modification of the space-time geometry, naturally comes together with the wave itself. Does that mean that masses are byproducts of electromagnetic waves? As far as free-waves are concerned the answer is negative. The perturbation in this case is negligible and non localized (since the wave develops at the speed of light). There is a relatively mild curvature of the space, which is not enough to produce durable tangible effects. The only interesting *geodesics* are straight-lines and are exactly those the wave is following. This is in agreement with the concept of free-wave; the modification of the created geometry does not influence its path, establishing no constraints on its motion. However, when waves start to interfere and build confined structures, then gravitational effects are not only present in the geometrical sense, but they show up in mechanical fashion. Thus, our next step is to examine what happens to free-waves when they interact.

1.7   Constrained waves

A wave is *constrained* when, due to external perturbations, it is obliged to leave its free path. This may also happen at the encounter of two or more photons; if the trajectories of all their points do not meet, the photons have no chance to influence each other; but, if there is an intersection of their supports, mutual constraints develop. Hence, during this interaction time, the waves are not of free type. It took me a while in order to understand how to convert into formulas the mechanism of such an impact. I then realized that a full theory was readily available. It is a question of geometrical deformations, so the theory of general relativity was there to help me. More drastically, I now believe that general relativity exists primarily to answer the problem of wave interactions, and that its role in gravitation is only incidental. The importance of this statement will become more relevant going on with my exposition. Thus, the suspicious reader is invited to stay calm a bit longer.

Collectively, the colliding photons locally perturb the geometric environment, following at the same time its new geodesics. Therefore, Einstein's equation is the right instrument to proceed with the analysis. The challenge is to translate the geometrical approach into the language of partial differential equations. It bears reminding that photon-photon scattering in vacuum is predicted by quantum theories and the effects of photon collisions are observed at high energies, resulting for instance in the production of electron-positron pairs[40]. In real life, the phenomenon is negligible and not energetic enough to be realized in laser laboratories. The continuous creation of "virtual" particles during subatomic electromagnetic scattering looks to an outsider a pure fantasy, an elegant way to say that nobody knows what really happens during those interplays, aggravated by the fact that quantum theories deny any other form of interpretation by masking everything under the mantle of the *uncertainty principle*. From the quantitative viewpoint, *Quantum Electrodynamics* is quite a precise theory, so that, the current shadowy assumptions must hide a more credible core. The understanding of these situations is then crucial for the development of my theory at a subnuclear level.

In the process of finding an extension to the differential model, one has to keep in mind that everything must be chiefly described through electromagnetic fields. However, in terms of energy, something new happens. At each point of the 3D space we assign a triplet $(\mathbf{E}, \mathbf{B}, \mathbf{V})$ (as before, for the sake of simplicity, we do not use here the quadri-vector formulation). This means that, at the intersection of their supports, two photons have to share their fields, which in most situations leads to a reduction of their global electromagnetic energy. Therefore, if one believes in energy preservation properties, the missing part has to be found somewhere else. Indeed, if a change of trajectory occurs, an acceleration is acting on the field $\mathbf{V}$, that, if we are lucky, is going to derive from a certain potential energy. The total energy turns out to be the sum of a pure electromagnetic contribution and a kinetic one. In terms of 4×4 tensors (those to be placed at the right-hand side of Einstein's equation), the first contribution is the electromagnetic stress $U_{\alpha\beta}$ (see sections 1.5 and 1.6) and the second one is the so called *mass tensor* $M_{\alpha\beta}$ (see (94) in appendix G). At least at an informal level, a link between electromagnetism and masses starts emerging. Some ideas in this direction have been also tried in [Woodside(2004)].

Thus, at the encounter of two photons, we find a significant modification of the geometrical setting. Differently from the case of free-waves, this

---

[40]From [Jackson (1975)], p.11: "There is a *quantum-mechanical nonlinearity* of electromagnetic fields that arises because the uncertainty principle permits the momentary creation of an electron-positron pair by two photons and the subsequent disappearance of the pair with the emission of two different photons [...] This process is called the scattering of light by light".

change actually involves both space and time, since one has to take into account that the velocity field **V** is now part of the game, both with an active role (that of driving the electromagnetic signal) and a passive one (**V** is influenced by the way the electromagnetic signal is developing). The use of the techniques of general relativity is now unavoidable[41]. Such a nontrivial curvature of the geometry is associated with a complicated metric tensor describing (at least locally) the space-time. This could be in principle put in relation to the vague presence of some mass distribution (this aspect will be better focused as going ahead with the exposition). The interaction of photons becomes in this way similar to that of massive bodies; they are like elastic fluid drops displaying electric and magnetic properties. We are still far from the creation of an effective mass, but the basis for a discussion has been set up.

In constrained waves, the triplet $(\mathbf{E}, \mathbf{B}, \mathbf{V})$ is no more orthogonal (see also footnote 16). Models in which the electric field may assume a longitudinal component are quite common, especially when examining the interactions with matter (see, e.g.: [van der Merwel and Garuccio(1994)], [Hively and Giakos(2012)]). As I noticed in the previous section, the local metric and the displacement of the triplets are strictly related, and this seems to remain true in more complex situations. In such a peculiar way, the electromagnetic fields provide both the information on the signal carried and the geometric properties of the space they are embedded in. The non-orthogonality of the electric and magnetic fields with respect to the direction of motion is also observed when a wave passes through matter, due for instance to *dielectric polarization*. We have no matter here (at the moment), but we are starting to detect the first signs of its presence.

In terms of governing equations, the differential model looks a bit more complicated than the one related to free-waves. Indeed, the free-waves model is included in the general one, as bodies in inertial motion are naturally expressed in Newtonian mechanics. The general set of equations combines the Maxwell's superstructure and the Euler's model for compressible inviscid fluids (see appendix G). The result is the description of an "immaterial" flow that carries vector quantities representing electricity and

---

[41] In [Einstein(1920)], the author writes: "A curvature of rays of light can only take place when the velocity of propagation of light varies with position. Now we might think that as a consequence of this, the special theory of relativity and with it the whole theory of relativity would be laid in the dust. But in reality this is not the case. We can only conclude that the special theory of relativity cannot claim an unlimited domain of validity; its results hold only so long as we are able to disregard the influences of gravitational fields on the phenomena (e.g. of light)".

magnetism. I think that this is, to some extent, what J.C. Maxwell had in mind when writing his papers, trying for example to interpret the attraction of magnets as the difference of pressure of certain micro-vortices resident in matter[42]. Indeed, in those works there is a constant reference to the properties of electromagnetism, viewed as a flow of some nature. Nowadays, this mechanical approach looks an ingenuously intricate gadget[43], but the reader will discover that Maxwell's intuitions were not that far from what I consider to be reality.

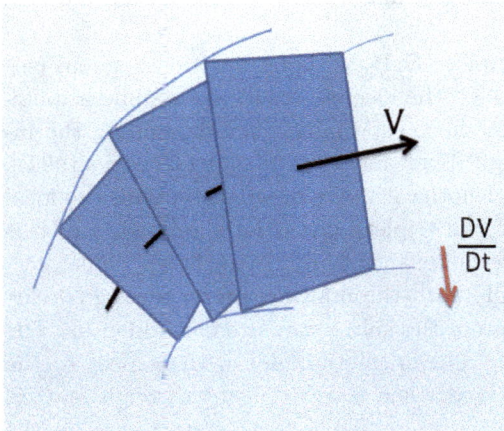

Fig. 1.4 *Wave-fronts subjected to a change of orientation during their development. An acceleration is acting on the vector **V** giving rise to a mutation of the trajectories of the rays. The rules of geometrical optics are broken in the flat space, but they survive in a suitable curved metric space where geodesics look bent.*

---

[42]In [Maxwell(1861)], the author writes: "Let us now suppose that the phenomena of magnetism depend on the existence of a tension in the direction of the lines of force, combined with a hydrostatic pressure; or in other words, a pressure greater in the equatorial than in the axial direction; the next question is, what mechanical explanation can be given of this inequality of pressure in a fluid or mobile medium? The explanation that most readily occurs to the mind is that the excess of pressure in the equatorial direction arises from the centrifugal force of vortices or eddies in the medium having their axes in the direction parallel to the lines of force".

[43]In [Maxwell Commemorative Booklet(1999)], F.J. Dyson writes: "The scientists of that time, including Maxwell himself, tried to picture fields as mechanical structures composed of a multitude of little wheels and vortices extending throughout space. These structures were supposed to carry the mechanical stresses that electric and magnetic fields transmitted between electric charges and currents. To make the fields satisfy Maxwell's equations, the system of wheels and vortices had to be extremely complicated".

Quantitatively, the acceleration is provided by the term $D\mathbf{V}/Dt$ (see figure 1.4 and (88) in appendix G). In fluid dynamics, the operator D/Dt is the so-called *substantial* or *material* derivative, describing the time rate of change of a quantity along a velocity field, as measured by an observer in motion following the streamlines. Geometrically, $D\mathbf{V}/Dt$ is linked to the concept of *parallel transport* of a vector, that leads to the definition of geodesics; the equation $D\mathbf{V}/Dt = 0$ models an autonomous system of infinitesimal particles, free from external perturbations and only subject to natural geometrical constraints. For a free photon embedded in a flat space, this means that the wave is correctly allowed to proceed undisturbed along a straight path. Nontrivial cases happen when $D\mathbf{V}/Dt \neq 0$; where some force is acting on the system; geometrically there could be the possibility that the space-time displays some curvature. Formalizing these ideas is a matter for experts and requires a level of abstraction that is not in my background. It has to be remarked that most of the work was done in the past century. General relativity is made on purpose to blend mechanics and geometry together, so I will just rely on known techniques, since they can still teach us a lot.

A new entry appears in my model: it is the pressure p (considered to be zero in the case of free-waves). In truth, dimensionally speaking, p is not a real pressure, but the problem is solved with multiplication by a dimensional constant. The presence of the scalar quantity p is an important indicator that something in the gravitational environment is going on. The introduction of p should be somehow expected: the so called radiation pressure is for example exerted upon surfaces exposed to luminous waves (see [Ashkin(1969)]); this punctuates its role in the interaction of light and matter. The possibility of applying the present model to the study of the acceleration imparted to a conductive obstacle by an electromagnetic signal is examined in [Funaro and Kashdan(2015)].

There are basically two new equations in the system (see appendix G). The first one replaces the condition $\mathbf{E} + \mathbf{V} \times \mathbf{B} = 0$ by the Euler's equation, where now the expression $\mu(\mathbf{E} + \mathbf{V} \times \mathbf{B})$ takes the meaning of "forcing term". Here $\mu > 0$ is a fixed constant, whose role will be clarified later on. When the wave is free such a forcing term is zero, so that pressure remains zero and no acceleration is acting on the velocity field. As a consequence the wave stays free. The same equation has another important physical explanation: it is the counterpart of the *Lorentz law*, in the case where only fields are involved. Light rays can be seen as the trajectories of infinitesimal charged particles, traveling at speed $\mathbf{V}$, under the action of the fields $(\mathbf{E}, \mathbf{B})$. Breaking the relation $\mathbf{E} + \mathbf{V} \times \mathbf{B} = 0$ amounts to applying a force to these objects, causing the deviation from their path. On the other hand, the term $\mathbf{E} + \mathbf{V} \times \mathbf{B}$ also appears when operating a Lorentz transformation to the electromagnetic tensor $F^{\alpha\beta}$ (see (72) and similar expressions). This indicates that such a term is implicitly present in the description of the wave without the necessity of introducing infinitesimal particles carrying mass and charge. The constant μ is the ratio charge/mass of these hypothetical point-wise particles. I propose μ to be approximately equal to 2.85×10^{11} Coulomb/Kg, for the reasons that will be explained in the sequel. This number is relatively small, with the effect of producing on radio waves extremely mild accelerations, i.e., not appreciable at a short distance. It is however well calibrated when one starts investigating atomic structures.

The last new equation participating in the system is a straightforward consequence of energy preservation and describes how pressure can raise (see (90) in appendix G). This effect comes with the non-orthogonality of $\mathbf{E}$ and $\mathbf{V}$. The extreme case where $\mathbf{E}$ is parallel to $\mathbf{V}$ (perfect longitudinal wave) is also handled by the equations (see (106)). The entire set (see also page 18) looks a potpourri of well-known physical ruling equations; we can actually recognize Ampère's, Faraday's and Lorentz's laws, blended into a mechanical evolutive setting. It has to be remarked however that everything is written in terms of pure fields and no effective matter is involved. By an inspection in [Clauser(1960)], p.84, or [Jackson(1975)], p.491, one finds a strong affinity with the model here studied. There the equations simulate the flux of charged particles (electrons) having a certain density distribution n. This similarity certifies that what I am doing here is not far from consolidated guidelines, except for replacing particles with fields, which is something unseen before.

This final version of the modeling equations furnishes primordial constitutive laws regulating light interactions. I will use these to build up matter, without the help of any further support. In this fashion, it will be interesting to rediscover how mechanical bodies behave and react to electromagnetism. There will be no big surprises, because both originate from the same roots. The sign of the electromagnetic energy tensor that, added to the mass tensor form the right-hand side of Einstein's equation, is the same as that corresponding to free-waves (see section 1.5). The two tensors may now balance each other, leading to interesting equilibrium configurations. From the practical viewpoint, the energy tensor $T_{\alpha\beta}$ is now defined as in

(94). The general set of modelling equations turns out to be compatible with relation (39), i.e.: $\partial_\alpha T^{\alpha\beta} = 0$.

Let me give some more technical instances. By taking the *trace* of Einstein's tensor equation, one arrives at a scalar equation, saying that the *scalar curvature* R of the space-time is proportional to p (see (97) in appendix G). In order to get this result, one must assume a suitable hypothesis on the mass tensor, which is equivalent to requiring the validity of the eikonal equation in the modified metric space (see section 1.4 and (96) in appendix G). In other words, if for an external observer embedded in a flat space the development of the wave-fronts of a constrained wave does not follow the rules of geometrical optics, the anomaly should not be noticed by the local observer traveling with the wave. When time elapses of a quantity δt, the fronts uniformly move a distance $\delta x = c\delta t$ along the orthogonal direction; the difference with the classical case is that the concepts of distance and orthogonality have to be referred to the local metric and do not necessarily coincide with the notion of the external observer (see figure 1.4). Finally, let me note that the scalar curvature does not provide accurate knowledge of the metric space, but when it is different from zero indicates that a serious deformation is going on; then, it is not by chance that $R \neq 0$ derives from the presence of a non-vanishing pressure. Curved space-time, deformed geodesics and pressure are various symptoms of a common pathology. For an inflexible mathematician, I am probably not using the proper rigorous language to express these ideas, but the important thing here is setting up the main framework.

Let me remark once again that for free-waves one has $p = 0$ and $R = 0$. The passage of a photon that, as said in section 1.6 produces a local shifting of the space-time, is not sufficient to create more severe distortions than those necessary to its motion. The pressure term is however ready to come out when a photon hits an obstacle. In the encounter of two photons, things can become really complicated. Certainly, at a very local level, the wave-fronts start following strange paths, but in most circumstances the definition itself of wave-front loses meaning, since these surfaces may not exist in global form. One has also to take into account possible changes of polarization twisting the fields, accompanied by breakage and reconstitution of topological settings. The classification of these occurrences may be largely variegated, thus the situation shown in figure 1.4 is oversimplified. More likely, in photon-photon interactions, the electric field at each point varies its angle with $\mathbf{V}$, with the possibility of turning around in the longitudinal direction[44]. Thanks to (90) pressure changes in time, but not with a monotone behavior and not uniformly in space. The whole reaction is highly nonlinear and resembles whirling flows. The problem of finding solutions to the model equations in such complex cases can be approached numerically, however, due to the nonlinearity and the hyperbolicity of the system the task is not that easy. The overlapped region where the interaction occurs is a transition zone in which the mixture of various ingredients may give the impression that virtual entities are momentarily formed[45]. As we will have time to discuss later, these non permanent passages may end up with the formation of stable particles and anti-particles, but more plausibly they just constitute transitory stages that, though necessary for respecting energy and momentum conservation, dissolve as the impact ends. One may also try to recognize in this effervescing choreography the micro-vortices conjectured by Maxwell (see footnote 42). It may be argued that, according to a principle of linear superposition, photons should not interact, a property that is commonly given for granted. Nevertheless, under suitable conditions, photons may actually interconnect (see, e.g.: [Firstenberg et al.(2013)], [Liang et al.(2018)]), and these are situations where my set of equations effectively come into play.

It has to be pointed out that the density ρ (that can also assume negative values) appearing in the mass tensor is not exactly a density of mass, but continues to be a density of charge. The reasons of this construction will be better clarified in section 1.8. Let me anticipate that it is not enough to have $\rho \neq 0$ to "feel" the existence of a mass, but a right amount of pressure p must also develop. My strong claim is that, up to dimensional adjustment, p actually plays the role of *gravitational potential* (see section 2.6). Due to this statement, the model equations are sufficient to describe the

---

[44]From [Keller(2005)], p.93: "The photon is a concept of the free electromagnetic field which only has transverse degrees of freedom, and in the initial phase of the emission process, where the source is still electrodynamically active, one cannot rigorously speak of photons in the wave mechanical sense. In the period of time where the particle(s) and field are coupled the electromagnetic field in general has both longitudinal and transverse parts, and with the spatial localization problem for photons in mind it is of interest to follow the space-time development of the transverse field (in the given inertial frame). As the field-matter interaction process is brought to a conclusion the transverse field smoothly approaches the field accompanying the just (newly) generated photon".

[45]From [Jackson(1975)], p.13: "In the atomic and subatomic domain there are small quantum-mechanical nonlinear effects whose origins are in the coupling between charged particles and the electromagnetic field. They modify the interactions between charged particles and cause interactions between electromagnetic fields even if physical particles are absent".

combination of electromagnetism and gravitation, without passing through the resolution of Einstein's equation and the full knowledge of the metric. The computation of the triplets $(\mathbf{E}, \mathbf{B}, \mathbf{V})$ and the scalar p provide enough information for a complete understanding of both electromagnetic and gravitational-like quantities. If we want these results to acquire a general validity, independently of the system of coordinates and the observer, one must write down the new set of equations in covariant form (as done in the case of free-waves). This goal does not present major difficulties, basically because it uses standard tools of general relativity (see appendix G).

My model comes from considerations based on energy and momentum preservation. Hence, in the impact of two waves we can recognize the same conservation rules valid for mechanical bodies. What remains at the very end of such close encounters continues to develop as free-waves would do, and no further accelerations are registered. I will explain later that the universe is completely filled up with electromagnetic signals; therefore, assuming the existence of pure free-waves is utopistic. As a consequence, the general situation looks rather involved, perhaps more similar to a chaotic broth. What happens in stars is an "illuminating" example on how things can be knotty, but I expect a similar degree of complexity also in unsuspected regions, assumed erroneously to be empty. The situation is not however so desperate since very organized patterns may emerge and drive the surroundings into increasingly sophisticated shapes. It is my intention to examine attentively this building process in the course of the following chapters. At the moment, only very simplified structures can be analyzed with the required accuracy: the case of *vortex rings*.

1.8 Vortex rings

Vortex tori are the most peculiar substructures that fluids can admit. The flow is distributed inside a well determined doughnut-shaped domain and follows a rotatory movement around the major circumference, so that the toroid axis is a symmetry axis (see figure 1.6). Sections are not necessarily circles and the internal hole of the toroid can also degenerate into a segment (*Hill's spherical vortex*). An easy way to get vortex rings is by ejecting a fluid into another fluid from the hole of a nozzle at relatively low velocity. Once the ring is created, it displays great stability properties. It can last for long periods of time and tends to rebuild if cut into pieces (see for example figure 1.5). Two vortices may meet and adhere. They may start performing a kind of dance where, with an alternate sequence, one ring enters inside the other's hole and reemerges on the other side allowing the second ring to enter its own hole (leap-frogging). If the impact is more violent, they can break into many smaller stable rings (see the impressive images in [Lim and Nickels(1992)]). Maxwell's colleague Lord Kelvin was really amazed by this behavior to the point where he thought that atoms (at that time, the basic inseparable constituents of matter) had actually a vortex shape[46] (see also [Silliman(1963)]). This was happening at the end of the nineteenth century, when knowledge of molecular composition was still at a primordial stage. A quick search on the Internet shows, starting from smoke rings, an impressive collection of videos and experiments concerning these amazing structures (see also [Van Dyke(1982)]). Here is an incomplete list of papers dealing with vortices and similar phenomena: [Maxworthy(1972)], [Pullin(1979)], [Shariff and Leonard(1992)], [Wakelin and Riley(1997)], [Elcrat, Fornberg and Miller(2001)], [Linden and Turner(2001)], [Sullivan et al.(2008)]. An astonishing property of vortex rings is that they seem very mildly affected by gravitation or other external uniform forces, so that they can travel horizontally almost undisturbed. Fine particles may be trapped in them and carried away without being influenced by gravity (see [Domon, Ishihara and Watanabe(2000)]). Such a "screening" effect is curious; a partial answer will be attempted later (see sections 2.2 and 2.6).

The connection with electromagnetism is imminent. It is possible to predict the existence of ring type formations (or more complex knotted shapes, according to [Kedia et al.(2013)]), inside which two or more photons are rotating in stable equilibrium. Such constrained waves behave exactly as material rings. In fact, from the mechanical viewpoint, they are ruled by Euler's equation. Moreover, they display electrical properties. Roughly speaking, the situation is similar to that corresponding to a wire, supplied by an alternating current, wrapped around a toroid surface. The wavelength of the current is a sub-multiple of the circumference section, so that the artifact is a complex resonant cavity. The wave is so imprisoned in a region that turns out to have electric and magnetic properties. There are however no physical boundaries, but the wave is self-induced to remain within its natural border. This is also true in the case of material fluids where an explanation relies on the balance between centrifugal forces and

---

[46]From [Thomson(1867)]: "After noticing Helmholtz's admirable discovery of the law of vortex motion in a perfect liquid – that is, in a fluid perfectly destitute of viscosity (or fluid friction) – the author said that this discovery inevitably suggests the idea that Helmholtz's rings are the only true atoms".

surface tensions. The issue of stability is not trivial as it seems and will be developed in section 2.2. In order to permit the wave to undergo such a localization, a suitable deformation of the space-time geometry is strictly necessary. Thus, the object also displays "gravitational" properties.

Fig. 1.5 *By puffing air into water, dolphins create wonderful ring-shaped bubbles. Successively, they play with these structures boisterously. Indeed, such vortices can persist for a long time and support extreme perturbations.*

Vortex rings of electromagnetic type are observed in nature. Let me mention a few examples. First of all, toroid structures are examined in *plasma physics*[47] (see, e.g.: [Chen, Pakter and Seward(2001)], [Hsueh et al.(2007)], [Barenghi and Donnelly(2009)]), where vortices may appear up to the nanoscale ([Gomez et al.(2014)]). Afterwards, *soliton rings* are commonly produced in optical laboratories by sending laser beams through a crystal lattice. In these studies, waves are supposed to carry angular momentum with the result of creating peculiar structures when propagating in nonlinear periodic arrays. We return on this in section 2.1 with more references. Rings are also found in astrophysics (see, e.g.: [Bryce, Balick and Meaburn(1994)]).

Finally, the atmospheric phenomenon referred to as *ball lightning* (see, e.g.: [Stenhoff(2002)], [Nikitin(2004)], [Bychkov(2014)]), where some

---

[47] In [Hazeltine and Meiss(2003)], p.107, about plasma confinement we can read: "A magnetic surface, or flux surface, is a smooth surface whose normal is everywhere orthogonal to the magnetic field. Thus the field lines lie everywhere in the surface; the generic flux surface is densely covered by a single line. When the confining field never vanishes and never intersects a material wall, each flux surface must be a topological torus".

form of radiation remains imprisoned for relatively short periods of time in confined errant regions, could also originate from electromagnetic waves trapped in rings[48] (see [Dawson and Jones(1969)], [Endean(1976)], [Alanakyan(1994)], [Rañada and Trueba(1996)]; and [Funaro(2018a)] regarding my viewpoint on the subject). Other light phenomena in atmosphere are reported for instance in [Zou(1995)]. In [Hasimoto(1972)], an amazing relation between the motion of a thin vortex ring filament and the nonlinear Schrödinger equation has been found. Such a kind of link is very important, since, in the sequel of this exposition, I will have a chance to focus on many aspects inherent in the role of electromagnetic tori in the composition of atoms and molecules[49].

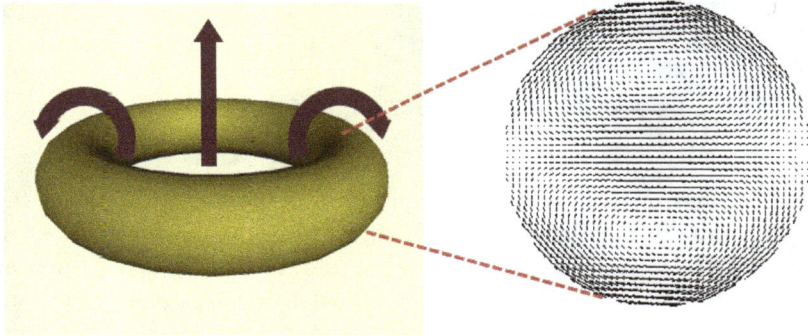

Fig. 1.6   *Photons rotate inside a toroid according to the arrows. The electric field lays entirely on each single section, while the magnetic field, orthogonal to it, oscillates back and forth parallel to the major circumference. This is an exact solution of the whole set of Maxwell's equations (see also the picture on the left in figure 4.4).*

---

[48] From _Wooding(1963)]: "Various explanations of these phenomena have been put forward, and I now suggest that the lightning ball consists of plasma vortex rings. Vortex rings are hydro-dynamically stable and are easily produced, for example, a smoke ring. [...] The vorticity of the lighting ball could be produced impulsively by asymmetric expansion of a lightning stroke, particularly if the stroke occurred in the vicinity of a solid object, having an aperture. Reports of ball lightning describe its arrival down chimneys or through doorway".

[49] From [Dragoman and Dragoman(2004)], p.196: "Note that solitons are the stable non-spreading NLS (NonLinear Schrödinger) solutions in one dimension, the stable features for three dimensions being the vortices".

Let me better explain what a toroid wave looks like. I am able to do this since exact solutions are available (see appendix G), at least in the case when the major diameter of the toroid is reasonably larger than the one corresponding to the section. With this hypothesis the section is practically a circle (see [Chinosi, Della Croce and Funaro(2010)]). These solutions appear to be the sum of two parts: a time-dependent one and a stationary one. The latter is going to be examined later. The time-dependent component is due to the circular motion of one or more photons displaying a sinusoidal modulation with respect to the rotation angle. The electric field lays on the section of the toroid, while the magnetic field, orthogonal to the first one, is oriented as the major circumference and circulates around the ring axis (see figure 1.6). The velocity vector field $\mathbf{V}$, orthogonal to $\mathbf{B}$ but not to $\mathbf{E}$, also belongs to the toroid's sections and indicates the direction of movement.

If one looks at the evolution of the electric field for the pure rotating part, several conclusions can be drawn. First of all, the condition $\rho = \text{div}\mathbf{E} = 0$ is permanently satisfied. Hence, we are solving the equations in the Maxwellian case. There are plenty of these waves, representing the actual solution space of Maxwell's equations, although these rotating photons can be considered anomalous in the classical sense (see section 1.1 and footnote 8). Contrary to free-waves, they do not propagate and their support is a doubly connected region of the 3D space (note that the toroid is exactly this). In this way, it is possible to simultaneously enforce all the conditions: $\text{div}\mathbf{E} = 0$, $\text{div}\mathbf{B} = 0$ and $\mathbf{E} \perp \mathbf{B}$. As I mentioned above, the electric field is not orthogonal to the velocity field, which is a typical situation characterizing constrained waves. By the way, having $\rho = 0$ and setting $p = 0$, there is no need to introduce the field $\mathbf{V}$, since the whole set of equations will be satisfied anyway. Indeed, both Euler's equation (87) and equation (90), containing the time derivative of p, become of the form $0 = 0$. Hence $\mathbf{V}$ can be totally arbitrary, a property that makes these solutions very permeable to the passage of free-waves. In fact, in cases where the parameter μ can be considered negligible, Maxwellian and propagating waves sum almost linearly. This is a very important observation because it reveals that behind complex nonlinear ruling equations one may find a linear response at specific regimes.

The solution (101) in appendix G and its derivations are very important, being the prototype of an immense variety of Maxwellian solutions constituting the support of real matter. Other, more involved, examples of periodic type are found in [Funaro(2014)]. For an external observer at rest, at any point inside the toroid, the electric field is turning around, following elliptic type patterns (possibly degenerating into segments). It is possible to individuate the surfaces enveloping the electromagnetic field at a fixed time (see figure 1.7); these are toroid-like. During the evolution they self cross and the principles of geometrical optics in the classical sense are completely disregarded. A similar situation is encountered for the Hertzian wave (see section 1.3 and figure 4.3 in appendix E). These are just examples of how a constrained wave can be amazingly unconventional. Let me remind the reader that, at the moment, I am only analyzing Maxwellian type waves, so that I am not adding anything new to what should already be known. The real revolution is noting that standard propagating waves do not belong to the solution space of Maxwell's equations, that includes instead many unexpected representatives.

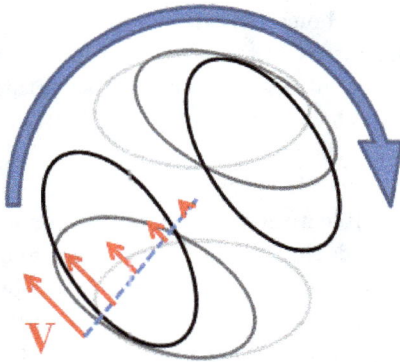

Fig. 1.7 *Evolution of the section of a surface enveloping the electromagnetic field in the case of the toroid solution of figure 1.6. A velocity field* **V**, *linearly increasing in intensity from the center can be defined. The magnetic field* **B** *is orthogonal to the page.*

The definition of wave-fronts is now getting problematic. A reasonable assumption is that these are the surfaces orthogonal to the velocity flow-field **V** of figure 1.7. In this case, the carried electric message has a component in the direction of motion. The limiting case is at the outer surface of the vortex ring, where **E** and **V** are lined up and **B** is zero; a setting very similar to that shown at the end of appendix G. Differently from free-waves, we do not have now the coincidence of the wave-fronts with the surface envelope of the carried fields. The global average of the electric field, taken over the toroid region in a period of oscillation, turns out to be zero. The same is true for the orthogonal magnetic field, that oscillates back and forth in the direction parallel to the major toroid circumference. Thus, although this localized electromagnetic entity has energy and (a latent) momentum of inertia, we cannot actually claim that it is a "charged particle". Such a spinning object is hard to detect in experiments, due to the fact that it cannot be accelerated under the action of electric or magnetic fields. It could be however decisive to maintain certain energy balances, and this is the only way to catch it. In fact, let us suppose that in an experiment some quantity that has to be preserved is instead missing; if we know that such a quantity has not escaped from the site in other forms, it can be well hidden in a transitory toroid storage area, as the one described above.

As anticipated at the beginning of this section, compatibly with the model equations, it is possible to add a stationary solution to the dynamical one, in such a way the result is still defined in the 3D toroid region. According to the new set of modeling equations, it turns out that $\mathbf{E}$ must be of radial type and growing linearly from each section's center (see figure 4.4). Again, the magnetic field $\mathbf{B}$ is orthogonal to the toroid sections (see figure 1.8). Now, it is compulsory to define $\mathbf{V}$, so that the right-handed triplet $(\mathbf{E}, \mathbf{B}, \mathbf{V})$ is clearly individuated. The angular speed of rotation ends up being constant, so that $\mathbf{V}$ grows linearly with the distance from the section's center (figure 1.7). It is very important to notice that there are no other displacements of the stationary orthogonal triplet $(\mathbf{E}, \mathbf{B}, \mathbf{V})$, except for the one indicated above. This says that the ruling equations are very accommodating as far as time-dependent solutions are concerned, but very strict regarding the stationary ones. Interestingly, the distribution of an oscillating electric field inside a solenoidal coil is still a matter of debate and, only recently, accurate experimental results have been made available (see [Lee, Lee and Yu(2005)]). The existence of stationary type fields and the possibility of superposing them on dynamical fields will allow us to introduce for instance the concepts of charge and magnetic momentum, otherwise unavailable in the Maxwellian case without permitting the solutions to exhibit singularities. I will return to these issues in the next chapter.

Note also that the property that light rays travel at the speed of light is not true anymore. In fact, this would be correct if we treated the photon as a point-wise particle globally moving at speed c. But, assuming that a photon has an internal structure, we have to recognize that, during a change of trajectory, there are parts of it traveling at different velocities, greater or smaller than c, with an average equal to c. Note that the relation $|\mathbf{V}| = c$ is not going to be true as far as constrained waves are concerned. This observation points out the limit of the theory of special relativity, that has to be replaced by general relativity (see also footnote 32). Working with the tensor structure of the metric it should be possible to adjust the clock transversally to the direction of motion, in such a way that the curving trajectory of any internal point of the photon moves at speed c, even if, during a certain amount of time, some paths look longer or shorter. Of course, talking about "short" or "long" has no absolute meaning and depends on the observer. In terms of the metric tensor the correction of the clocks involves the activation of the matrix coefficients linking space and time variables.

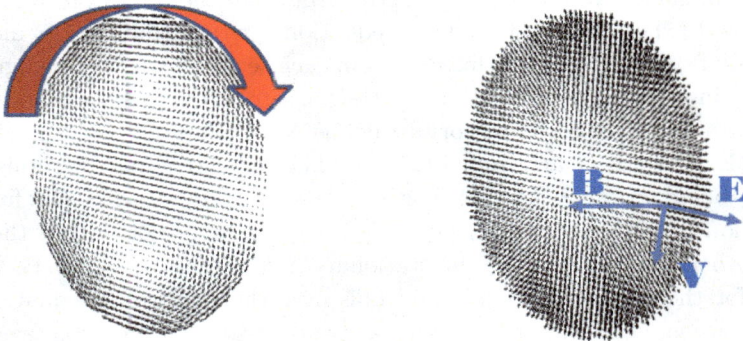

Fig. 1.8 *Stationary orthogonal triplets may be added to the rotating fields. In the first picture (see also figure 1.6) the whole section containing the electric field rotates according to the arrow. In the second picture, compatibly with the model equations, a stationary radial electric field has been added (see figure 4.4). Analogously, an orthogonal stationary magnetic field has also been summed, in such a way that the triplet ($\mathbf{E}, \mathbf{B}, \mathbf{V}$) is right-handed.*

A pressure $p \neq 0$ now comes into play. Roughly speaking, p (negative or positive) signifies a rarefaction or compression of the wave-fronts, that, for an observer at rest, do not develop according to the rules of geometrical optics. To be more precise, the evolution of the wave-fronts does not follow the *Huygens principle* in the flat geometry; however this should rather be true in the metric space modified by the wave itself, according to Einstein's equation. In truth, in section 1.7, I specified that, although not visible in the model equations, the eikonal equation is available to show its validity.

When riding a rotating photon (i.e., from the viewpoint of an observer placed in the modified metric), based on the considerations developed in the case of free-waves, I expect the following to happen. Independently of the size of the photon, the wave-fronts are planes of infinite extent marching parallel, one next to the other. There is no gradient of pressure in the azimuthal direction, so the fronts seem to satisfy the rules of geometrical optics. In any case, the radial gradient of p should be different from zero with p proportional to the scalar curvature of the space-time (see (97) in appendix G). In this way, the photon's observer may feel the effect of a transverse acceleration that can be interpreted as a gravitational field. I will show in the next chapter how to confer a mass to the whole system. Let me confess that I do not have precise confirmation of what is written in this paragraph, since a rigorous analysis can only follow after solving Einstein's equation with the goal of finding the geometrical structure inside the toroid. Such computation looks at the moment to be rather involved.

A question arises: do we really need to solve Einstein's complicated equation in order to have a clear picture of what happens during photon-photon interactions? In sections 1.6 and 1.7, I noticed that the magnitude and the orientation of the electromagnetic field provide sufficient description of the geometric evolution. In other words, the triplet $(\mathbf{E}, \mathbf{B}, \mathbf{V})$ is a kind of surrogate of the local metric tensor. If this was true, knowledge of the fields $(\mathbf{E}, \mathbf{B}, \mathbf{V})$ should already be enough for a complete analysis. It is however necessary to include as further information the scalar quantity p. Since the set of equations modelling the evolution of the fields has validity independently of the geometrical environment (covariance), a suitable description of our phenomena is already well represented in the flat space. In such a particular reference domain, the equations speak the language of classical physics, so that one may try to interpret, without the help of fancy generalizations, what is behind the electromagnetic vortex structure. Nevertheless, as it will emerge in the continuation of my exposition, the flat space viewpoint is a very rough averaging of a complex reality made of a gigantic number of time-developing pieces, suitably patched together, each one representing a micro-universe.

There are a couple of ways to describe in classical terms the stable rotation of two photons, and this is the key point to comprehend the unification process between electromagnetic and gravitational forces. We can first attribute a density of charge ρ and a density of mass ρ_m to the toroid solution (see (98) in appendix G). From the computations (in the flat space) it turns out that the density of charge is uniformly constant, while the density of mass is strictly related to p (which approximately varies as a quadratic expression of the electromagnetic field). We know that a charge travelling in a combination of electric and magnetic fields can be put in steady rotation (the *cyclotron* exercise). This is due to the action of the Lorentz force, that in some form also appears in the modelling equations. On the other hand, a mass can be put in orbit in a gravitational field. Without having to deal with actual masses or charges, but only with fields, the equations already incorporate all these features. The evolving structure has to be seen as a unique body, inside of which relations between fields are built according to the same identical laws ruling point-wise charges and masses. The model equations for fields are the germs from which, once matter has been finally built, one recovers the classical physics laws.

I believe that this is the completion of the work initiated by Maxwell himself, i.e., to translate into the jargon of fields what can be observed in nature. The core of my reasoning is the following. If I am able to construct something that has all the features of an electron (the most simple piece of matter) by constraining photons into a toroid, and I can do it with model equations inspired by standard electromagnetism and fluid mechanics, then I have to expect that classical physics will descend as a consequence. Somebody, at this point may argue that classical physics is only part of the game, and that, particularly in the case of objects of the size of an electron, quantum mechanics is prevailing. I am not bothered by this criticism, since I will also show how to recover quantum properties of matter. In truth, my story started in the opposite way (see section 3.8). I was searching for a justification of quantum phenomena and I ended up dealing with electromagnetism. Therefore, the most interesting part has still to come and I will face quantum problems without the need to alter the model equations.

What I just said is correct only when we add the stationary part to the rotating one. By the way, in my theory there are no pure stationary solutions. Hence, there are no charged bodies in the way it is usually supposed. As we shall see in the next chapter, an electron is quite a complicated object, that, only to the first approximation and at a certain distance, can be roughly described as a point-wise elementary charge, emanating electric field in accordance with Coulomb's law (note that in this case the corresponding potential tends to become singular when approaching the charge). According to my interpretation, an electron is made of rotating photons, trapped in a toroidal region, therefore, it is a dynamical object. The time-evolving part is needed to create the topological structure, the added stationary part provides for the effective mass and charge. Nevertheless, the stationary part cannot survive without the dynamical one; it would degenerate into singularities as Coulomb's formula prescribes. Note instead that the non-stationary part can exist with or without the stationary one, giving rise to different types of elementary structures. With these observations, I have prepared the ground for the next chapter, where more sophisticated solutions displaying strong similarities with subatomic particles will be studied.

Chapter 2

The Subatomic Environment

Quello che noi ci immaginiamo
bisogna che sia o una delle cose già vedute,
o un composto di cose o di parti delle cose altra volta vedute
Galileo Galilei, scientist

2.1 Electrons

I start in this chapter my adventure towards a systematic description of matter and its properties. We already have an elementary particle: the photon, and we know almost everything about it, at least in the case when it behaves as a free-wave. Finding place for light-quanta within the context of a classical theory of electromagnetism is a fundamental achievement, considering that the problem has been on the priority list of open questions in physics for decades[50]. The initial erroneous assumption was to consider Maxwell's model immutable[51]. We overcame this difficulty, so it is possible to restart the research process from the very beginning and see how the clarification of the wave-particle duality enigma can bring benefit to us.

---

[50] From [Planck(1927)]: "For the elementary quantum of action plays no part in Maxwell's equations. From dimensional considerations it would be entirely impossible to introduce this quantity into Maxwell's equations unless additional constants should appear therein. [...] Again and again the question arises, must we really ascribe to the light-quanta a physical reality, or is there after all a way of taking account of them, which preserves the validity of Maxwell's classical electrodynamics?".

[51] From [Planck(1927)]: "You all know that theoretical physics, which developed and progressed consistently through two hundreds years, and only a generation ago seemed near to its final conclusion, has now entered a critical period, fraught with serious consequences. Not all its fundamental principles have been questioned! For its more general and at the same time its simplest laws, such as [...], and the Fundamental Equations of the Electromagnetic Field, are just the ones which have so far withstood successfully the most severe trials, and serve now as ever before as the guides for wider exploration".

In section 1.8, I provided instructions for building other distinctive objects, constituting the basic bricks for further developments. In the simplest form, these entities are composed of photons self-constrained in bounded regions of the 3D space, having a toroid-like shape. Let me remind the reader that, according to the way I introduced free-waves, it is not compulsory for a photon to be a small object, but any size may fit the definition (see section 1.2). Nevertheless, physicists are used to dealing with photons by imagining them as microscopic entities associated to quantized phenomena. I will show the reader how to handle this discrepancy[52]. For the moment, let me specify that quantum properties of photons are not inherent to photons themselves, but to the way they interact with matter. Hence, before discussing these connections, we must discover what exactly is "matter". For this reason, it is necessary to start from the simplest levels.

I begin by examining the case of the electron, since, according to my equations, full explicit computations can be carried out (see appendix H). In particle physics, the electron is not supposed to have internal structure. It carries a unit of charge and a very small mass. Moreover, it also displays *spin*. The concept of spinning particle was vaguely proposed by W. Pauli and later reconsidered by R. Kronig and G. Uhlenbeck. At the beginning, this property was associated with an actual rotation of the massive body, represented by a tiny charged sphere. Lately, it was made clear that this was not quantitatively compatible with the electron's characterizing assumptions. Connections with the concepts of *parity* and *helicity* are usually established when talking about spin. I am more precise on these issues in section 2.3. Nowadays, the idea of spin is purely based on mathematical hypotheses and fits experimental data quite well. Physicists do not bother anymore comparing spin to classical mechanical behavior, so the real reason for its existence remains an indefinite philosophical exercise[53].

---

[52]From [Feynman(1985)]: "We know that light is made of particles because we can take a very sensitive instrument that makes clicks when light shines on it, and if the light gets dimmer, the clicks remain just as loud - there are just fewer of them. Thus light is something like raindrops - each lump of light is called a photon - and if the light is all one color, all the raindrops are the same size". Concluding from this sentence that diaphanous low-frequency photons are almost uncolored large raindrops would be too quick; the explanation is going to be more delicate and necessitates more understanding about the way matter is structured.

[53]From [Dragoman and Dragoman(2004)], p.2: "... many physicists consider quantum-classical analogies as a mere curiosity, without any physical implications. Classical physics is often regarded as a rudimentary and superseded theory without much relevance for the more refined and modern quantum theory, although any outcome of a quantum process is registered by classical instruments located in the classical world".

As far as I am concerned, the electron is composed of confined spinning photons and this confers an intrinsic dynamical behavior to it. The form assumed is the simplest one: sinusoidal photons constrained in a vortex ring. According to the spinning verse, I will assign an arrow to the toroid axis, that it will be improperly called *spin axis*. A possible section configuration (the most elementary) is visible in figure 4.4, where one recognizes two symmetric oval patterns; they rotate around the axis orthogonal to the page and placed at the center of the disk. A frequency can be associated to this object depending on its size and on the fact that the internal information shifts along closed circular-type orbits with an averaged speed *c*. This sets an inverse proportional relation between the frequency and the minor diameter of the toroid. It is interesting to observe that, while the major diameter turns out to be fixed due to stability arguments (I will return later to this issue), there is no prescribed size for the minor diameter, so that there is a continuum of frequencies that can be associated with an electron (see figure 2.1). In other words, there is no single representative of the family of electrons, since the ratio between the minor and the major diameters may vary, preserving however charge and mass. Note also that the movement of a vortex ring is not comparable to the rotation of a rigid 3D body. In the latter case an absolute concept of spin cannot be introduced, since the revolving velocity depends on the observer's framework. The above considerations are the results of the study of exact solutions defined on toroid shaped domains.

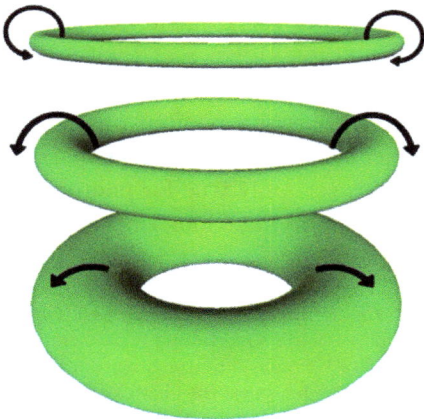

Fig. 2.1 *Three different aspects of an electron built from spinning photons. Mass, charge and major diameter are fixed. Frequencies, inversely proportional to the minor diameters, may vary with continuity, but the kinetic energy as well as the electromagnetic one remain unchanged (see appendix H).*

Spin is not however enough to characterize an electron, since mass and charge have still to be taken into consideration. They derive from the stationary component, mentioned in section 1.8, that is, in first approximation, linearly added to the evolving part. Remember that this operation is compatible with the whole set of equations presented in appendix G. In particular, the dynamical part is a Maxwellian wave ($\rho = 0$ and $\mathbf{V}$ undefined), but the addition of the stationary part requires the solution of the entire set of model equations. In particular, the current term of the Ampère law (that in the case of free-waves has been associated with a sort of immaterial charge density traveling within the support of the wave), is now witnessing the presence of an effective charge, with density $\rho \neq 0$, trapped inside the toroid. I also specified that solitary propagating waves are not compatible with the assumption $\rho = 0$. In other words, free photons carry with them the potentiality to generate charge, which is actually manifested when they are forced to remain in bounded regions. Thus, stationary fields overlapping the spinning photons immediately provide the particle with electric and magnetic properties, while to understand how mass is created we need more investigation.

There are a few possible ways to add stationary contributions. I examine here one possibility and I will add more explanations later in section 2.3, where *antimatter* is introduced. Extraordinarily, there are not so many degrees of freedom to work with. I consider this to be a positive fact, since in the end the property of being an electron is very exclusive and cannot be shared with counterfeit copies. Without the steady part, we get a spinning, uncharged, massless particle that may agree with the definition of *neutrino*, i.e., another remarkable particle, one of the primary ingredients of nuclear physics. Neutrinos of standard particle physics represent a very reduced subset (carrying possibly an extremely small mass), whereas I will use the term to denote structures that are far more general and unconventional.

Let me specify here that a controversial question is often raised when dealing with massless neutrinos; it is in fact claimed that zero-mass particles must travel at the speed of light[54]. If this is certainly true for photons, the observation does not apply to the barycenter of neutrinos that, as far as

---

[54]For example, in [Okun(1987)] p.11, we find: "The concept of helicity is not Lorentz invariant if the particle mass is non-zero. The helicity of such a particle depends upon the motion of the observer's frame of reference. For example, it will change sign if we try to catch up with the particle at speed above this velocity. Overtaking a particle is the more difficult, the higher its velocity, so that helicity becomes a better quantum number as velocity increases. It is an exact quantum number for massless particles (neutrinos and photons)".

I am concerned, may assume whatever velocity the circumstances require. Indeed, they are a sort of storage container of energy and momentum, aimed to preserve mechanical rules. The mistake comes from attributing a rotatory motion of the particle around an axis, but the situation is different for the toroid, where, for example, it is hard to find a trivial system of coordinates where the object is at rest.

In the course of this exposition, we will have time to appreciate all the features of neutrinos. Contrary to the electron, these enjoy a lot of freedom and may adapt themselves to an unlimited number of configurations (see also [Funaro(2014)]). Their equilibrium is sometimes precarious; in my version, they can degenerate into a photon and, in this form, travel at the speed of light, ready to restore their initial configuration as the circumstances are favorable. Arbitrary stationary fields with $\rho = 0$ may be added to a neutrino. For the sake of a full rehabilitation of old-fashioned electromagnetism, it is essential to remark that, in toroid form, these are Maxwellian waves. They belong to a picturesque and variegated space, having nothing in common with propagating signals emitted for example by an antenna device.

The geometric properties of the space-time must be invoked in order to understand the existence of spinning objects (see, e.g.: [Ricci and Ruggiero(2004)]). Regarding photons, let me briefly review a few facts already discussed in the previous chapter. The rotating photons, being energy carriers, modify the geometric nature of the environment. The corresponding metric tensor, may be computed from Einstein's equation, after plugging the appropriate energy tensor on the right-hand side. In this context, as far as an observer from a flat space is concerned, the geodesics are not straight-lines. Hence, the trajectories of the rays are bent. This impression is not shared with the observer riding the photons, who, due to the metric modification, is convinced of proceeding straight. Now, we have a chance to get a stable situation when the path followed by the photons perfectly coincides with the self generated geodesics. Put in other terms, energy and momenta associated with the photons are able to ply the space and produce a toroidal geometric structure where the geodesics are closed paths; the same paths are exactly those the photons are actually following and the process is somehow locked. I was not able to compute exact solutions of Einstein's equation in such a complex case, therefore the above considerations are not formally proven.

Similar configurations have already been studied in the general relativity framework, although at a qualitative level[55], with applications in context of the theory of *black holes*. Roughly speaking, the situation is similar to the one of twin planets rotating in equilibrium around a common center. If speeds and distance are appropriately set, and there is no emission of gravitational waves, the configuration is stable. The coupled system modifies the space-time geometry creating curved geodesics, and, at the same time conform to such pre-determined trajectories. This is in the spirit of Einstein's interpretation of gravity, to which we have to include here the additional difficulty of dealing with electromagnetic fields.

A crucial aspect contributing to the stability of a neutrino-like solution concerns the sign of the electromagnetic stress tensor $U_{\alpha\beta}$ to be placed on the right-hand side of Einstein's equation (see (56) in appendix E). As observed in section 1.6, I opted for a switch, subverting the standard set up (see also the justifications provided in [Funaro(2008)], p.92). It is nice however to discover that the modification is also meaningful with respect to well-established solutions. There is in fact a version of the *Reissner-Nordström metric* (see (63) in appendix E) that suitably matches the sign flip. The metric represents a black-hole. The usual version is affected by a technical intrinsic limitation that influences the minimal radius of the black-hole, whereas my updated version eliminates such a restriction, allowing for the existence of black-holes also of the size and the mass of the electron. This is a further symptom of the correctness and consistency of my proposal to alter some conventions.

In physics, the term "neutrino" denotes very specific representatives of the family I am considering here, therefore I expect some perplexed reactions to my statements. It is better then to face the explanation of the role of neutrinos when the reader is more acquainted with my theory. Therefore, let us go ahead with the analysis of the electron. Very few gravitational interactions end up with masses in fixed orbits. Similarly, not all the interactions of two photons lead to stable periodic configurations. In this regard, we have to admit that the formation of an electron is a very rare event. One

---

[55] From [Wheeler(1955)]: "On the basis of classical general relativity as it already exists, and without any call on quantum theory, it turns out to be possible to construct an entity that we call geon. This object serves as a classical singularity free, exemplar of the "bodies" of classical physics. Of such entities there exist in principle a great variety, distinguished from one another by mass, intrinsic angular momentum, and other properties. The simplest variety is most easily visualized as a standing electromagnetic wave, or beam of light, bent into a closed circular toroid of high energy concentration. It is held in form by the gravitational attraction of the mass associated with the field energy itself".

needs the right shape and the right energy of the colliding entities. In most cases, the final product consists instead of smashed free-waves diffusing all around. It has to be kept in mind however that, independently of the event taking place, energy and momentum conservation must hold. This rule is implicitly contained in the balance laws representing Einstein's equation.

At this point, it is necessary to give more details about the stationary part of the electron's solution. According to figure 1.8, the electric field **E** is radially distributed on the toroid sections, in such a way that $\rho \neq 0$ is nearly constant (when the ratio between the sizes of the minor and the major diameters is small). In practice, **E** grows linearly from the section's center. For the reasons detailed in [Funaro(2008)], p.122, one must have $\rho > 0$; therefore, the electron turns out to be a <u>positive</u> uniformly charged particle, despite the convention of considering its charge negative. I hope this information is not going to bring consternation to the reader, who is referred for an explanation to section 2.3; everything is however under control (just in case, I would suggest to wearing the oxygen mask and breathing normally).

The volume integral $e = \epsilon_0 \int \rho$, where e is the electron's charge and ϵ_0 the dielectric constant in vacuum, quantitatively allows for the determination of ρ (that amounts to fixing γ_0 in (101), appendix G). The velocity field **V** is orthogonal to the stationary **E** and establishes the sense of rotation of the dynamical part (see figure 1.7). Note that, when summing up the steady and the time-evolving contributions, one loses the orthogonality relation between **E** and **V**. The stationary magnetic field **B** remains orthogonal to both **E** and **V**, forming a right-handed triplet; thus, it lies on closed loops around the toroid axis. This confers magnetic properties to the particle, a very important feature that will be reconsidered many times in the course of this exposition. There are no singularities (**E** and **B** are bounded inside the electron's domain), contrary to what we would have in the case of a point-wise particle. Finally, an expression for the pressure p is also available (see (102) in appendix G). I remark that the displacement of the fields examined in appendix F refers to a cylindrical setting, but the whole machinery can be adapted to rings (see [Chinosi, Della Croce and Funaro(2010)]). It is very interesting to observe that the displacement of the stationary **E** is analogous to that obtainable inside a homogeneous dielectric (see figure 2.2). This shows that our electron more and more resembles a real piece of matter, though it only consists of pure fields.

I will start discussing questions regarding mass from section 2.2, arriving at the nitty gritty in section 2.6. For the moment, let me just give some technical points. Based on the definition given in [Funaro(2009b)], a relativistic density of mass can be derived by taking the M_{00} component of the mass tensor; up to dimensional constants, this is proportional to the scalar $\rho c^2 - p$ (see (98) in appendix G). Therefore, the mass m of a particle is a suitable multiple of the volume integral of the density of mass; more in detail one has: $m = (\epsilon_0/\mu) \int (\rho - p/c^2)$, where $\mu \approx 2.85 \times 10^{11}$ is the same constant appearing in the model equations (see (89) in appendix G). The integral is extended to the 3D domain enclosing the solution. For example, as far as a single free-photon is concerned, one has $p = 0$ and $\int \rho = 0$, hence the formula yields $m = 0$, in agreement with the fact that photons carry energy but display no mass. A similar conclusion holds for the dynamical part of the electron. This time we have $p = 0$ and $\rho = 0$, showing that neutrinos have zero mass. In the case of the full solution (i.e., including the stationary fields), m exactly agrees with the electron's mass. In accordance with special relativity m is not an invariant quantity. To tell the truth, μ, which is approximately of the same magnitude as the elementary charge e divided by the electron mass m, has been set up with the purpose of obtaining this result. This means that, once the constant μ is appropriately defined, it is possible to construct solutions of the entire set of model equations, such that they have the same global size, charge and mass as an electron. We should not forget however that μ must remain constant, so that we are not allowed to modify it in the future. Therefore, a significant validation of my theory may follow by proving that, with the same identical μ, other charged massive particles can be built with characteristics similar to the originals. Let me honestly confess that, due to the complication of the problem, I cannot rely at the moment upon such a result.

In the end, in my view, an electron is a cloud of spinning fields, displaying all the necessary quantitative and qualitative properties. It is a piece of "matter" that cannot be divided further. Indeed, the object has to be taken as a whole, since its parts do not survive independently, but are only sustained by a delicate equilibrium process. This construction is quite challenging, but can be supported by some experimental observations. For instance, once an electron is formed, it is very hard to destroy it; anyway, if by chance it interacts with a *positron* (the *anti-particle* of the electron, see figures 2.3 and 2.4), the couple breaks up into pieces, that are in most cases high-frequency free photons. This exactly shows the nature of the two particles. Put together, particle and anti-particle form an unstable structure and give back the substance they are made of, i.e., electromagnetic fields.

The massive part disappears due to the annihilation of the respective stationary components and to the breakage of the geometric environment (this is a crucial point that will be examined later). Conversely, one needs relatively large powers in order to generate electrons from photons. However, from the collision of high-energy photons (*two-photons* physics), realized in accelerators, it is possible to create the conditions for (indirect) particle generation. These phenomena are well documented and suitably coded in *Feynman diagrams*, covering a wide range of cases. These magical recipes can predict with extreme accuracy the outcomes of almost any reaction. What is really lacking in this taxonomy is an explanation that goes beyond the schematic approach and the consolidated rules of quantum electrodynamics. And this is what I am trying to do in this chapter.

Based on my computations, it turns out that an electron may assume a multitude of different shapes, sharing however the same physical properties (see figure 2.1 and appendix H). All of them have a toroid shape, more or less of the same size. The major diameter is practically fixed around 2.7×10^{-15} meters (2.7 fm), in accordance with nuclear dimensions, but the minor diameter may vary. Thus, the ring could be very thin, so that the frequency of rotation of the photons is extremely high. Conversely, the ring can be fat and the frequency low (above 10^{22} Hertz, however). Probably, there exists a natural asset, corresponding to a certain choice of the minor diameter that optimizes some parameters, but its determination seems not to be easy without more serious numerical computations. All the members of this family have the same charge and mass. They are actually different representations of the same particle in various possible "states", described by a parameter (the frequency, for example) that may vary with continuity. The fact that the frequency is not uniquely determined is not a problem; on the contrary, it provides the electron with an incomparable adaptability. From experiments, the volume of an electron turns out to be very small, at the point that its size is considered to be negligible. I think that this is instead due to the thinness of its support and to its high versatility, making the particle hard to catch. I also would like to specify that the supporting ring is not rigid; in fact, I expect the electrons to be very deformable when subjected to external perturbations, exactly as fluid vortex rings behave in their various manifestations, both in nature and laboratories (see the caption of figure 1.5). Thus, electrons can be squeezed and twisted, showing a resistance that can only be broken through the interaction with antimatter.

At first sight, it may sound strange that objects of different volumes display the same mass and charge. Of course, this is because the respective densities scale inversely with volume, but I believe there is a deeper reason, especially concerning electric charge. Incidentally, in the framework of *differential cohomology* (a branch of algebraic geometry), charge quantization is strictly referred to rings. As a matter of fact, cutting out a ring from the 3D space amounts to drilling a real hole in it. Smooth closed paths can be divided into two categories: those that can be shrunk to a point and those interlaced with the ring, that cannot be reduced to a point without cuts. Note that a spherical hollow does not produce such a dichotomy, so it is not a real hole. In topology, the magnitude of the ring has no influence, since the discipline is only concerned with the study of shapes. Therefore, in the presence of a toroidal hole we have a quantum of charge, independently of actual size. It has to be noted that the value of the charge is computed through a line integral on a closed path taking also into account the displacement of the electric field. It turns out that, although displaying a toroidal shape, neutrinos have charge equal to zero as is expected to occur. These properties have been checked using exact solutions, but there must be a way to reach general conclusions by directly working on the differential problem, and analyzing its underlying geometric nature. This would also prove that, provided ruptures do not occur, certain quantities continue to be preserved even under severe alterations of the electron's support, a result that analytically cannot be inferred, due to the unavailability of more complex explicit solutions. A lot of material is already available for Maxwell's equations, so that such an investigation would start from good credentials. I am certainly not an expert on this subject, therefore I leave the question to those skilled.

Let me mention a trivial, but less known, experiment concerning with magnetic fields and the topological properties of closed lines of force (see [Leedskalnin(1945)], although the explanations given there are a bit naive). Take two simply-connected pieces of (non magnetized) iron, shaped in such a way to form an annular body when joined together (it is enough that their union has a hole). Insert a conductive wire into the hole and let an electric current pass through it for a fraction of time, inducing magnetic field to "circulate" inside the iron. Remove the wire. The two pieces are now solidly glued together. The effect is long lasting and it takes a good amount of strength to tear the pieces apart. When divided, the two parts do not reconnect anymore, because the rupture of the toroid-shaped object has passed through a topological transition, completely modifying its magnetic status. Moreover, some energy that remained stored, even for a long time, when the toroid was intact, is ready to come out in other forms after the breakage. These observations are aimed to add further confirmation for the exceptional resilience properties of the electron ring.

The idea that matter is of electromagnetic origin is not so weird. I am not indeed the first one claiming that photons are constituent of elementary particles[56], also considering that such a statement, at least in the case of the electron, looks to be an obvious consequence of real experiments. Anyway, most of the attempts in this direction did not completely succeed in their goals. They started with good initial assumptions, but they failed to get final convincing answers. As far as I am concerned, all in all the problem is always reducible to the chronic incapability of casting photons within the classical electromagnetic context[57]. My additional contribution consisted in proposing the appropriate modeling equations and showing, from this improved standpoint, that creation of matter from fields is theoretically achievable through classical tools, implying ultimately that fields are actually the sole ingredient. The nonlinear properties of the ruling equations are fundamental in this analysis[58].

An interesting consequence of my reasoning is the difficulty (and perhaps the impossibility) of building *magnetic monopoles*. Maxwell's equations in vacuum do not actually distinguish among electric and magnetic fields; the role of the two entities can be exchanged, without altering the general setting. As a matter of fact, one could build neutrino type solutions ($\rho = 0$)

---

[56] From [Miller(1984)], p.33: "... These were the chief developments that led Wilhelm Wien to propose research toward an 'electromagnetic basis for mechanics', that is, an electromagnetic world-picture, as an alternative to the apparently sterile inverse research effort toward a mechanical world-picture. In the electromagnetic world-picture, mechanics and then all of physical theory would be derivable from Lorentz's electromagnetic theory. A far-reaching implication of this research program was that the electron's mass originates in its own electromagnetic field quantities as a self-induction effect".

[57] Regarding Einstein's determination to find an unification theory of electromagnetism and general relativity, in [Gross(2005)] it is written: "The core of his program was to include electromagnetism and derive the existence of matter in the form of, what we call today, solitons. As Einstein understood, nonlinear equations can posses regular solutions that describe lumps of energy that do not dissipate. Thus one could start with the nonlinear field equations of general relativity and find localized particles. [...] As far as I can tell, Einstein knew of no example of solitons or any toy model that exhibited his hopes".

[58] In studying the possibility of obtaining electrons from confined photons, the following statement is reported in [Williamson and van der Mark(1997)]: "Although circulating solutions of the linear Maxwell equations have been shown to exist, the fact that the electron does not have arbitrary mass means that some extra, presumably non-linear, effect must also play a role".

where electric and magnetic fields are switched. By dropping the condition $\rho = \mathrm{div}\mathbf{E} = 0$, as done in the case of my equations, one can enlarge the space of solutions, but the differentiation between magnetic and electric components becomes extremely neat. An electron type solution where such fields are interchanged is now not admissible. Moreover, the presence of the term $\mathbf{E} + \mathbf{V} \times \mathbf{B}$ emphasizes the asymmetric role of the two field flavors. I tried in [Funaro(2008)] to introduce a more general set of equations allowing for the possibility to also include the case $\mathrm{div}\mathbf{B} \neq 0$, but I do not think this is a viable alternative, since it causes many other fundamental properties to be lost.

Let me add at this point some historical notes. In the so-called *classical model*, the electron is supposed to be a deformable spherical distribution of charge. Its assemblage requires the help of a certain external pressure of not well specified nature. A first study of this structure was conducted by H. Poincaré[59]. The model was abandoned due to the difficulty of explaining many open questions. The *Parson magneton* is a toroid simulation of an electron, where a collection of infinitesimal "charges" are circulating along continuous paths wrapped around a ring. A magnetic field is generated inside the body. The structure was the first one to incorporate charge, spin and magnetic moment in a single unitary object. This is a naive version of what I actually got. Nevertheless, according to modern physics, particles must remain under the competence of quantum mechanics, thus frustrating all the efforts of finding explanations via classical tools. Nowadays, the common trend is to attribute point-wise properties to the electron. Not satisfied of such a rigid construction[60], various communities of researchers have manifested their alternative ideas, through books and conferences. This material deserves attention, though it diverges from the path I am following. Conscious that a lot of bibliography is available, I

---

[59]Concerning the existence of some hypothetical pressure maintaining an electron in shape, we can find in [Poincaré(1906)]: "J'ai cherché à déterminer cette force, j'ai trouvé qu'elle peut être assimilée à une pression extérieure constante, agissant sur l'électron déformable et compressible, et dont le travail est proportionnel aux variations du volume de cet électron. Si alors l'inertie de la matière était exclusivement d'origine électromagnétique, comme on l'admet généralment depuis l'expérience de Kaufmann, et qu'à part cette pression constante dont je vien de parler, toutes les forces soient d'origine électromagnétique, le postulat de relativité peut être établi en toute rigueur".

[60]From [Bostick(1985)]: "But the author wishes to enter his philosophical dissent against the continuation of the unqualified and unquestioningly accepted 'portrait' of this mathematical-point electron that is presented in texts an treatises on quantum mechanics. This catechism about the intrinsic nature of the electron has already been perpetuated for more than two generations".

limit my list to the following authors: [Dirac(1967)], [Bostick(1985)], [Filipponi(1985)], [Mac Gregor(1992)], [Williamson and van der Mark(1997)], [Arcos and Pereira(2004)], [Simulik(2005)].

When I started to look for rotating photon solutions, the first idea was addressed to spherical configurations. I realized that the goal was not achievable, and I moved my analysis to the investigation of rings. It was amazing to discover fluid vortices and their exceptional stability properties. It was even more surprising to find out the above mentioned topological properties of electrically charged rings, their relations with Maxwell's equations and the natural disposition of Maxwellian solutions to lie on doubly connected regions. This confirmed that I was advancing in the right direction. The theory that I am presenting here does not disclose untried fields of research, but has the advantage of setting up the links between various classical milestones. Past scientific advances brought a rich harvest, which is now ready to be used. Electromagnetism had still some "little bugs"; however, when properly fixed, is one of the pillars making the entire edifice solid (see footnote 51).

Rotating light is present in a multitude of applications under the keywords: *whispering galleries, photonic crystals.* Whispering gallery resonators consist of electromagnetic waves trapped in cavities (originally the name was referred to acoustic phenomena), smoothly guided to circulate around by continuous reflection, and returning at the origin with the initial phase. Spherical, cylindrical and ring-shaped whispering galleries are commonly produced for a broad range of industrial applications. Typical areas of interest are in fiber telecommunications or biosensing. For a general review, the reader is directed to [Snyder and Love(1983)], [Oraevsky(2002)]. Due to the property of bending radio signals, the ionosphere itself can be seen as a whispering gallery (see [Budden, Martin and Mott(1962)]). A survey on optical confinement in microcavities, showing some typical nanofabrication technology resonators, can be found in [Benson et al.(2005)]. Further related papers, among many others, are: [Kruglov and Vlasov(1985)], [Deutsch, Chiao and Garrison(1992)], [Mabuchi and Kimble(1993)], [Tikhonenko et al.(1996)], [Firth and Skryabin(1997)], [Kevrekidis et al.(2001)], [Bigelow et al.(2004)], [Boriskina(2007)], [Efremidis et al.(2007)], [He, Malomed and Wang(2007)], [Shen and Fan(2007)], [Fisher et al.(2009)], [Boriskina(2010)], [Xiao et al.(2010)], [Kamor et al.(2012)], [Karzig et al.(2015)], [Piazza(2015)], [Papasimakis et al.(2016)].

Before ending this section, there is another important issue I would like to discuss. It concerns the electron's boundary. The solution I am considering here is a kind of bounded and isolated displacement of fields. This means that I am not taking care of what is happening outside the toroid region. My electron is indeed a charged particle, but, at the moment, it has no opportunities to transfer the property of being a charge to the surrounding space. The object ends at its boundary and does not interact with other entities. I will call this a *bare particle*. This is certainly not in agreement with the standard concept of charge, displaying an unbounded electric halo, quadratically decaying in magnitude at infinity. Therefore, if we do not want this particle candidate to remain just a nice mathematical exercise, it is necessary to face the problem of providing it with a suitable habitat. For a while, I will continue to neglect the existence of any organization of the fields far outside the bare particle and I will stay in the neighborhood of the ring region.

2.2   More on the electron structure

A crucial question is to study what happens on the particle boundary. First of all I must say that, although what I am going to show is intuitive and logical, I have no formal theoretical results yet. A more serious analysis of the modeling equations, combined with a good knowledge of fluid dynamics is needed. A preliminary analysis can be carried out on the "neutrino component", i.e., the rotating fields with $\rho = 0$. Explicit computations show that, at the boundary, the magnetic field is zero and the electric field oscillates back and forth along the direction of rotation determined by the vector $\mathbf{V}$. According to this behavior, we expect a deterioration of the metric at the boundary of the toroid. This might be also confirmed by the degeneracy of the triplets $(\mathbf{E}, \mathbf{B}, \mathbf{V})$ that collapse onto one-dimensional curves.

The particle's border is effectively a 2D transition zone, where the model equations should continue to hold, though in some degenerate form. Note that it is an infinite far-away "horizon" for an observer inside the ring (see also section 1.3). Instead, an observer placed on the boundary (if we think this could be somehow possible) would only be able to see in the direction of the wave motion along the surface. From our standard flat space, we can distinguish between an interior, a boundary and an exterior, and these three entities turn out to be independent, non-communicating universes. Across the barrier, it makes no sense to speak about discontinuities. The jump situated between the internal and the external worlds is only apparent; as a matter of fact it cannot be measured. Observers from both sides consider the border zone as an infinitely distant unreachable frontier. An observer in the transition zone has no notion of the direction joining the two separated parts. It has to be remarked that I am working in a purely inviscid context. Differently from viscous flows, there is no possibility for a single point of the fluid to communicate with the neighboring points, via the mechanism of diffusion.

The background setting is slightly modified allowing for ρ to be different from zero. In this case, the notions of charge, mass and radial pressure enter the scenario. Here, in absence of more rigorous explanations, due to the difficulty of handling toroid shapes, things start to be more dogmatic. Following standard arguments, it is reasonable to assume that the gradient of pressure normal to the boundary is zero, implying that the centrifugal acceleration is compensated by the electrodynamical one, exerted by a Lorentz type force (see (87) in appendix G). In such a situation we should get stability, and we do not need to go through the computation of the metric tensor to actually check it. This is the trick I used to find the electron's boundary (see appendix H). The neutrino part, which is more adaptable, serves as support for the entire object. As I anticipated in section 2.1, the balance is realized for a family of toroid configurations, displaying the same global size, charge and mass, but dissimilar regarding the frequencies of rotation (see figure 2.1). Moreover, these configurations are referred to the particle at rest, but I expect the main properties to hold true under the action of external perturbations. Experiments conducted with real fluids, even in the presence of viscosity, confirm the impressive resistance of vortex rings under the effects of heavy deformations. For the exact determination of the quantities involved in the construction of the electron, the role played by the constant μ appearing in the model equations is very important. Such a constant is dimensionally equivalent to a charge divided by mass, and its magnitude has strict influence on the *charge/mass* ratio of the electron.

At the edge of the electron, the peripheral speed of rotation can be measured to be more than three and a half times the speed of light. Somebody may argue that this violates the theory of special relativity. I do not see however any inconsistency. As argued in section 1.7, the theory of general relativity expressed in tensor form, can provide much more insight with respect to the restricted theory, from where only an averaged behavior can be recovered. For example, when a bullet travels at constant velocity, we mean that its barycenter is moving at that velocity. Usually, the bullet is also spinning around its axis, so that there are points on the bullet that proceed, in spiral motion, faster than the speed of the barycenter. This aspect can be detected through conservation laws that, together with energy, also take into account momenta. Similarly, in the analysis of electromagnetic waves, we can distinguish between a general evaluation, based on scalar energy conservation, and a deeper understanding, where the internal structure enters into consideration.

Let me spend a few lines formally expressing the above concepts. All free-waves develop at constant velocity c. This constant actually appears in the modeling equations (see (84) in appendix G), and establishes the link between the scale of measure of distances and that of time. For free-waves the electric field is c times the magnetic field and both are orthogonal to the direction of motion. When the wave is constrained the situation changes. The electromagnetic fields are not totally "exposed" during their motion, but they start developing a longitudinal component. An extreme case is the one where $\mathbf{B}$ is zero and $\mathbf{E}$ is lined up with $\mathbf{V}$. One can check that to build this solution there is no need to know the constant c; it just disappears from the equations. Therefore, a pure longitudinal wave of this type can in principle travel at any velocity. If $\mathbf{V}$ follows a straight path, explicit solutions of Einstein's equation are also available in such a critical situation (see [Funaro(2008)], section 5.1). If $\mathbf{B} = 0$, and $\mathbf{E}$ oscillates back and forth in the direction of motion $\mathbf{V}$ with a frequency ω, then the scalar curvature of the deformed space-time ends up to be proportional to ω^2.

This is also what is expected to happen at the neutrino's boundary, although exact computations in toroid-shaped regions are at present not available. Examining the solution found for a cylinder (see (101) in appendix G), in the rotation of one or more photons it is possible to identify internal trajectories where the information develops with the electromagnetic fields nearly transverse to the direction of motion, as in a very thin free-wave. Here the speed is not exactly equal to c because the condition $|\mathbf{E}| = |c\mathbf{B}|$ is not fulfilled. As far as more external (or internal) trajectories are concerned, the electric field starts having a component in the direction of motion (see figure 4.4). With this inclination, the rays may travel at speeds greater (or smaller) than c. This is true up to the boundary, where $\mathbf{B} = 0$; there the electric field is completely longitudinal and could shift in principle at any velocity. But, at this point, the metric looses control on the radial direction and it makes no sense to go beyond such a limit. According to the equations (see (90) in appendix G), breaking the orthogonality between $\mathbf{E}$ and $\mathbf{V}$ generates pressure (positive or negative), which is an indicator of the fact that the wave is of constrained type. From the above

considerations, a very important remark reemerges. In the evolution of rotating waves there is a precise relation between frequency and maximum
radius allowed, due to the fact that the evolution occurs at an averaged
speed equal to c.

A subject of research in special relativity is the study of the so called
rigid rotating body. In particular, one may ask himself what happens to
the internal structure of a fast spinning disk when the points far from the
center rotate at speeds approaching that of light (or faster). Constrained
electromagnetic phenomena, as those introduced here, belong indeed to the
same class. The substantial difference is that the circulating points are
not totally independent observers, but they are all linked by the continuous electric field. Therefore, to carry out an analysis, the instruments of
general relativity have to be utilized in full. The entire structure develops
as a whole with constant angular velocity, without breaking the rules of
Einsteinian relativity. As I said, by going towards the edge, the electric
field gradually assumes a more pronounced longitudinal asset. This also
sets a limit to the magnitude of the body, exactly when the electric field
is completely longitudinal. The extremal velocity is greater than that of
light by an amount that also depends on the number of photons involved in
the rotatory movement (see (103) in appendix G). This shows that elegant
solutions of the rigid rotating body problem may emerge from the inclusion
of suitable electrical bonds between the points constituting the object; the
study gets far more complicated from the technical viewpoint, but more appropriate as far as its physical meaning is concerned. A deeper analysis may
suggest a review of the definitions of dielectric and permeability constants
as functions of the rotating framework (see [Pellegrini and Swift(1995)]).

Do the above considerations mean that it is possible to travel faster
than the speed of light? The answer is yes and no. Theoretically (and
surprisingly), Einstein's equation allows for solutions developing at any
speed and I also remarked that in the process involving the interaction of
photons there are parts where the information actually evolves in super-
luminous fashion. On the other hand, unperturbed electromagnetic waves
can only proceed at speed c and, when interacting, they can be faster
only in a controlled neighborhood. I do not know if, through an ingenious
amplification process, one could be able to launch objects at any speed. My
impression is that, although permitted by the theory, super velocities are in
general not present in nature, unless in an amount of time and space strictly
necessary to safeguard the rules of conservation of energy and momenta.

Let us now go back to the discussion of the electron, so that we also must consider the added fields having $\rho \neq 0$. At the boundary, three pieces of information are known. These concern with the electric field (orthogonal to the surface), the magnetic field (tangential to the surface) and pressure. For a ring with a small circular section, they are practically constant in magnitude. I went through the computations by assuming the pressure (see appendix H) to be zero at the central circumference inside the toroid (associated with the major diameter). I thought this was a reasonable choice, though other possibilities may be tried. With the idea of conferring stability to the whole body, my request was that the normal gradient of pressure had to be zero on the toroid surface. Thus, in the end one finds that a negative pressure is present on the electron's boundary (see (116) in appendix H). The pressure is also negative at the interior (radially decreasing), so that one can confer to p the role of a potential producing attractive forces (see section 2.6). On the boundary the pressure is constant in the case of a cylindrical geometry, or for a ring where the minor diameter is negligible with respect to the major diameter. In a more realistic case, p displays small variation in the direction of $\mathbf{V}$, depending on the curvature of the boundary surface. The phenomenon is due to a greater compression of the toroid's sections near the ring hole. In particular, there is a difference of pressure between the region of the ring surface near the toroid axis and that far from it.

If we now immerse the bare particle in a totally empty space, there are evident radial discontinuities across the boundary regarding $\mathbf{E}$, $\mathbf{B}$ and p. We need to fill this gap. Due to the *Magnus effect*, a fluid vortex ring immersed in another fluid moves along its axis, in such a way to create a constant flux through the hole. This is partly due to viscous effects[61]. Through such a mechanism, the rotating boundary of the ring transmits the motion to the outside fluid and forms another external larger and slower ring. The fluid dynamics of a ring vortex, inside and outside the well-defined toroid region, is quite complex, but extensively studied both theoretically and experimentally. Analogies can be noted with the development of a typhoon, where the fast spinning central kernel is usually surrounded by air, also in rotatory motion[62]. Sharp layers develop at the separation interface in the

---

[61] From [Batchelor(1967)], p.523: "From the theoretical point of view the striking property of all observed vortex rings in uniform fluid is the approximate steadiness of the motion relative to the ring when the ring is well clear of the generator. There is some decay of the motion always, presumably due to the action of viscosity, but the decay is less for larger rings, suggesting that the motion would be truly steady at infinite Reynolds number".

[62] Regarding the development of typhoons, one can read in [Kuo et al.(2004)]: "One way

viscous regime; hypothetically, they should become discontinuities in the inviscid case. The outside region is wider and its angular velocity is smaller than that of the core; this is also true due to momentum conservation arguments (large diameter, small frequency, and vice versa). As I will show you later (section 3.8), this coordination is the key point for understanding the quantum properties of matter. It might also be an explanation of why the *Planck constant h* is actually "constant".

When we put a bare electron in a sea of photon-like electromagnetic energy, we can expect a similar effect. In this case, there is no viscosity in the model, so it is not very clear how the flux through the hole may develop. Before guessing some explanations, I advise the reader that I am now dealing with pathologies that are still a matter of research in fluid dynamics. Therefore, I will only be able to speak informally. First of all, the particle is not a perfect rigid body and, through oscillations of the boundary around the equilibrium position, can impart acceleration to the surrounding electromagnetic waves. To this we have to add the above mentioned difference of pressure, somehow related to the local curvature properties of the surface. A further more significant hypothesis can be connected to the phenomenon of "adherence", developing between the two fluids separated by a steady interface, even if diffusive effects are not directly coded in the model. These circumstances create the conditions for an organized movement of energy in the immediate neighborhood of the bare particle. As a consequence, a set of photons, trapped in a larger external toroid-type domain, may start floating around the primary vortex ring. This new secondary ring might also degenerate into a *Hill's* type vortex, i.e., a "fat" toroid with the hole reduced to a segment. Other intermediate situations, inspired by real fluid motion, may also originate; however for the moment we will stay with the simplest case. Let me also point out (and this is an important remark) that the creation of structured whirlpools around electrons are not automatic, but strongly depend from what is happening in the environment, so that the taxonomy is extremely variegated.

Pictorially, a secondary spinning neutrino is embracing the bare electron and its existence and stability are somehow granted by the presence of the inner particle (see figure 2.2). Being a larger region and knowing that photons move on average at the speed of light, the new set displays a

---

to produce a halo of enhanced vorticity around an intense vortex is through a binary interaction in which the large, weak vortex is completely strained out. It will be shown that this mode of interaction is most likely to occur when the peak vorticity in the small, strong vortex is at least 6 times that of the large, weak vortex".

smaller frequency. Later, I will assume that such a frequency is inversely proportional to the distance from the source (mathematical details are given in appendix I). Inside and outside the separation surface, the operations can be conducted in total respect of the modeling equations. But, what really happens at the interface? I cannot provide a complete explanation because the question is very deep and touches the foundations of mechanics. On the other hand, if I want my model to be an exhaustive description of what is around us, all the most basic questions have to emerge and be dealt with. It is a good sign that they derive from the lowest level of the construction process, to later become the empirical laws observed in experiments. Let us look at this more deeply.

The bare electron is a kind of balloon, filled up of circulating fluid, that remains at rest in empty space. In particular, such a bubble does not explode. In classical mechanics, one may say that there exists a *surface tension* keeping the object well-balanced. Where does this tension come from? I cannot invoke the pre-existence of a membrane with an appropriate atomic character, since there are no suitable substructures. An explanation must be found within the context of the underlying geometrical structure. As I specified before, the pressure p turns out to be different from zero at the boundary.

In the study of *free-boundary flows*, in the case of *ideal fluids*, the difference of pressure across a surface boundary in equilibrium is proportional, via a certain constant σ, to a scalar quantity obtained by taking the divergence of the *normal vector field* with respect to the surface under consideration. To the first order of approximation, the above quantity may be put in relation to the so called *mean curvature*. In the case of the toroid, by denoting with R_1 and R_2 the minor and the major radii respectively, the mean curvature is on average equal to $\frac{1}{2}[1/R_1 + 1/R_2]$; more precisely, it is $\frac{1}{2}[1/R_1 + 1/(R_2 + R_1)]$ at the outmost equatorial circumference, and $\frac{1}{2}[1/R_1 + 1/(R_2 - R_1)]$ at the innermost one. Hence, the curved shape of the ring surface testifies to a difference of pressure between the interior and the exterior of the ring.

In practical applications, depending from the types of fluid involved, the constant σ is determined experimentally. Here σ should instead be deduced from the model, without the introduction of additional rules. This could be theoretically possible because the equations are not only describing the evolution of the single fluid velocity $\mathbf{V}$ (that by itself is probably not a sufficient condition to reconstruct the behavior at the boundary), but they follow the entire electrodynamical history (i.e., they are coupled with Maxwell's type equations), providing for suitable physical properties of the (infinitesimally thick) separation membrane. Information might be recovered by passing to the limit on the width of a suitable thin layer built around the surface separating the inner region from the outer universe. A possible starting point for a more accurate analysis is the *Bernoulli equation*, that can be easily recovered from the set of model equations (see (91) in appendix G). Note that Bernoulli's principle was devised on purpose to deal with changes of pressure caused by different speeds of a fluid. A typical application is the study of the transverse pressure in a *Venturi tube*, which is an effect very similar to the one we are considering here.

Another direct link between pressure and curvature is obtained as follows. By taking the trace of Einstein's equation, one discovers that p is proportional to the scalar curvature R (see section 1.5 and (97) in appendix G), which is an excellent geometrical invariant. In a two dimensional space R is the so-called *Gaussian curvature*, which is quite easy to compute with standard tools. In 4D (including time) the situation is more intricate (see ([Ricci and Ruggiero(2004)]). Now, it is not just the 3D shape of the toroid that decides the function R, but how things develop in time. Even if the set is apparently stationary, there are photons lapping the surface and they are associated with a frequency which is inversely proportional to the size of the toroid section. As we saw above, at experimental level at least, there is a straightforward nexus relating a suitable curvature of the 2D free surface and the pressure of the separated fluids. Some paragraphs above I also said that the scalar curvature is expected to be proportional to ω^2, where ω is related to the photon's frequency; this is the result obtainable in simplified circumstances, at least. I am tempted to deduce from this property that the difference of pressure across a separation boundary is proportional to the difference $\omega_1^2 - \omega_2^2$ of the squared frequencies. Compatible with this guess is the observation that the radial pressure present at the boundary of the bare electron actually behaves as the squared frequency (see (116) in appendix H). There is no direct relation however between R and the mean curvature, thus it is difficult to jump to conclusions. I am not experienced enough to be able to proceed with these arguments in order to find a correct place for all the pieces of this puzzle with the hope of more precise theoretical results. For the moment, it is enough to have recognized that a problem exists and its possible solution may rely on geometrical assumptions.

Always in accordance with the equations, a stationary electric field in gradient form, i.e.: $\mathbf{E} = -\nabla\Phi$, may be added to the dynamical solution corresponding to the outer secondary ring. By setting $\tilde{\rho} = \mathrm{div}\,\mathbf{E} = 0$, it is sufficient to demand that $\Delta\Phi = 0$ and impose boundary conditions in order to get a global continuous solution defined on the union of the two encapsulated rings. This can be actually done without introducing singularities; the opportunity comes from the presence of the hole in the secondary toroid, constituted by the primary vortex. The setting is exactly the same we are used to seeing in electrostatics. Indeed, the outer shell (up to its external boundary) simulates the electric field of the exterior of a charged homogeneous dielectric (see figure 2.2). In conclusion, one can construct a continuous prolongation of the stationary electric field outside the bare electron, in such a way that $\tilde{\rho} = 0$, whereas ρ is different from zero inside the particle. Both the stationary parts must also be sustained by a dynamical geometrical environment. In fact, the existence of the stationary components is subjected to the presence of the background system of rotating photons, that curve the space-time and actually provide for a stable support.

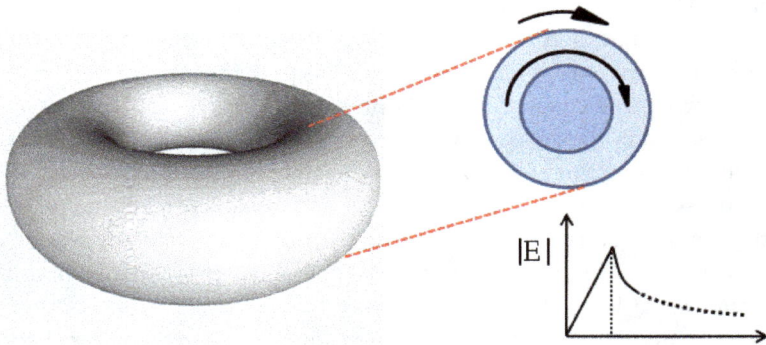

Fig. 2.2 *Two encapsulated spinning rings. The inner one is an electron dragging the outer one at lower frequency and causing a discontinuity in the velocity field. The stationary electric fields is described by a continuous function. In particular, the intensity of* **E** *grows linearly inside the electron ($\rho > 0$ constant) and decays outside, displaying zero divergence ($\tilde{\rho} = 0$). The illustration above is qualitative, hence it does not reflect the real shape of the secondary ring and its peripheral velocity.*

As far as the magnetic field is concerned, the situation is not that clear. The inner stationary component reaches the boundary tangentially, therefore, its prolongation outside might not be necessary. According to classical physics a toroidal solenoid supplied with constant current does not produce magnetic field at the exterior. It has to be taken into account however that a dynamical magnetic field (in symbiosis with the electric one) is always present. We are however discussing here very specific situations, simple enough to develop some mathematical speculations. Other displacements are available, when the outer fields start spiraling around the electron and things begin to be much more complicated if we allow the electron to deform its shape.

The results of this section can be recapitulated by saying that from different sides of the same frontier, the electromagnetic waves of the bare electron and the outer shell march at different velocities, according to suitable eikonal equations based on deformed space-time geometries. The geometrical setting degenerates at the common border. There are no measurable discontinuities for the field $\mathbf{V}$ across the 2D surface, since its differentiation is only taken along the streamlines. An analysis of the behavior of the model equations in the neighborhood of a separation boundary should guarantee a certain number of properties. Indeed, I would expect at least the continuity of the stationary component of the electric field across the boundary, as well as the continuity of the normal gradient of pressure. Note instead that the pressure itself may be discontinuous (only its gradient on the connected regions appears in the model equations). In the sequel, I will take these conditions for granted, although at the moment they are still a conjecture.

2.3   Antimatter

Before facing more complex issues, it is better at this point to introduce *antimatter*. Qualitatively, antiparticles are perfectly equal to their corresponding particles, except for a switch in the sign of the charge. In my view, the triplet $(\mathbf{E}, \mathbf{B}, \mathbf{V})$ is going to be replaced by $(-\mathbf{E}, \mathbf{B}, \mathbf{V})$. Accordingly, the model equations must be corrected by changing the sign in correspondence to the electric field. This means that we can obtain the same identical solutions we had before, where right-handed triplets are now replaced by left-handed ones. In this way, we can first get *anti-photons* and, successively, build antimatter. An all-inclusive set of modeling equations may be written by specifically placing the sign $+$ or $-$ in front of $\mathbf{E}$, following the orientation (right-handed or left-handed) of the current triplet. Formally, the classical Maxwell's equations only produce "right-handed waves".

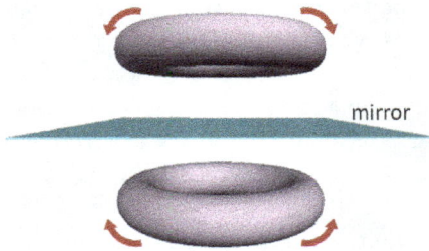

Fig. 2.3 *Mirror image of a vortex ring. After a rotation of 180 degrees, the object is super-imposable upon its reflection. The most basic neutrinos are almost of this kind. They are made of photons, while anti-neutrinos are made of anti-photons. Geometrically there is no distinction, but the internal constitution is slightly different.*

As we switch the sign of one of the two electromagnetic fields, the Poynting vector changes orientation and the wave moves in the opposite direction, still remaining right-handed. If we want Maxwell's equations to generate left-handed waves, one needs to modify for example the sign of the *curls* in Faraday's and Ampère's laws, obtaining an unusual version. The solutions of the modified system are specular images of the standard ones. The same situation is found for my model equations in the case of free-waves. These weird but meaningful remarks are not found in common treatises on electromagnetism, though I believe that in practice, the two versions cannot be distinguished until we remain within the domain of free-waves.

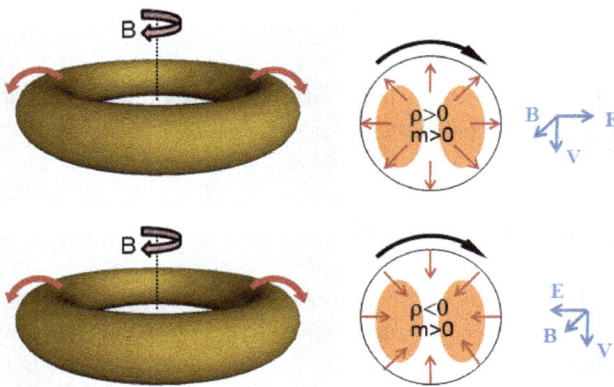

Fig. 2.4 *An electron (top) and a positron (bottom). The first one is right-handed, whereas the second one is left-handed. Except for the orientation of the electric field, they display identical properties.*

As electrons are made by rotating photons in a toroid region, their anti-particles (the *positrons*) are identically realized by using anti-photons, producing opposite charge but identical mass. The same can be said for other types of particles. Note that my electron is not the mirror image of a positron (see figures 2.3 and 2.4). In fact, depending on whether one analyzes the particle or its anti-particle, the electric field points outwards or inwards, and there is no way to pass from one displacement to the other through a reflection. The mirror image of a positron turns out to be an unstable particle; it corresponds indeed to the attempt of building a proton, that, as we shall see, has a much more complex form.

However, by removing the stationary part and leaving the neutrino time-dependent component, one gets an entity that at first sight coincides with the specular reflection of its anti-particle. The subtle difference is that the first one is made of photons and the latter of anti-photons. In this way, neutrinos and anti-neutrinos, although not perfectly coincident, may be related through a kind of *parity conjugation*. From theoretical arguments, the coincidence of a particle and its anti-particle implies that their mass is zero. The possibility of assigning a positive mass to pure neutrinos is indeed very remote, although this analysis is an active field of investigation. As far as I am concerned, neutrinos are generic photon aggregations confined in a region of space. They do not display any neat charge or magnetic momentum, since the fields are in constant movement with zero average over a period of oscillation. Based on my theory they have mass equal to zero (ρ is zero and consequently p is also zero). As a matter of fact, there is no way they can be accelerated by external factors. This remains true if we add a stationary component having ρ equal zero. If the stationary part is not of this type, neutrinos can partially acquire mass becoming potential particles. Electrons and protons (and their antiparticles), are examples where neutrinos become stable massive entities. According to my interpretation, there are infinite species of neutrinos, but only very few of them are effectively recognized by nuclear physics, because they are those principally involved in reactions. They rarely interact with matter. Regarding this fact, I will try to be more specific at the end this section.

An electron and a positron may form an unstable atom (the *positronium*) before collapsing and annihilating. In truth, what actually disappear are the electric stationary components, being equal and with opposite sign, while the neutrino parts merge in an unstable structure that decomposes into its constituents, i.e., pure photons. The transition is made with the respect to global energy and momentum balance. Note that an electron and a proton are not expected to annihilate just because they have different charges. In fact, it is important to also consider what happens to the stationary components of the magnetic field. In the electron-positron interaction, if the particles are properly oriented, the magnetic parts cancel out, while this is not true in the case of the couple electron-proton, that will be better studied in sections 3.1 and 3.2. There, we will also learn how spin orientation affects interactions[63]. The electron-positron collapse is a clear example of two "solid" pieces dissolving in gamma rays; what an incontrovertible evidence of the electromagnetic nature of matter (and antimatter)! Note that positrons in a universe made of matter tend to extinguish. This is not only true because of the violent interaction with electrons, but because they are eroded by the right-handed electromagnetic vacuum background (see chapter three). In my opinion, the reason why there is more matter than antimatter in our universe is based on the trivial observation that, if the two kinds were in perfect balance, we would not be here to speak about these concepts, since no significant durable structure would have emerged from the primitive setting[64].

Positrons (and anti-particles in general) can be momentarily created (before they are reabsorbed by matter) in various nuclear reactions. In my model, anti-waves are generated every time the triplet $(\mathbf{E}, \mathbf{B}, \mathbf{V})$ changes orientation. This may happen quite often, if the dynamics of the evolution allows for it. An interesting possible solution fully discussed in [Funaro(2008)], chapter 5, is when $\mathbf{B} = 0$ and $\mathbf{E}$ is lined up with $\mathbf{V}$. The orientation of the triplet is then undetermined. In this case, the radiation may reemerge in the form of wave or anti-wave, depending on the context. If the intensity of the fields is strong enough and the region sufficiently large, anti-waves can produce antiparticles and we can actually see them. Otherwise it is just a transition event that could not be recorded.

The term *chirality* is attributed to objects that are not super-imposable on their mirror image. This property is clearly noticeable in our everyday life from small molecules to living organisms, with very few excep-

---

[63]From [Fritzsch(1983)], p.38: "The difference in the lifetime of the two forms of positronium can now be explained more fully. In parapositronium, the electron and the positron annihilate each other and produce two photon quanta. The annihilation of orthopositronium can be described in a similar fashion; in this instance, however, three photon quanta are produced".

[64]"Doomed to vanish in the flickering light, disappearing to a darker night, doomed to vanish in a living death, living anti-matter, anti-breath"; from Pioneers Over C, Van Der Graaf Generator, *H To He, Who Am The Only One*, Charisma/Virgin Records (1970).

tions[65]. Nuclear experiments suggest that the micro-universe is chiral and left-handed. In chemistry, chiral molecules are very often encountered and they mostly show a left-handed (*levorotatory*) orientation. Typical naturally occurring amino acids are left-handed, and so are the corresponding proteins. This has a remarkable influence on the whole biological environment, at the level that right-handed molecules may be toxic or have devastating effects on living beings (I will return later in section 3.7 to the examination of some biological questions). The mystery of the occurrence of only one prevailing form of chiral molecules in nature has no direct explanation in chemistry, unless one start looking at atoms' nuclei. The discovery that β-particles emitted from radioactive nuclei do not display symmetric behavior, led scientists to hypothesize a sort of chirality also in the micro universe. Now, it is well accepted that most of the particles are left-handed and their antiparticles are instead right-handed. This perfectly fits my theory: it is enough to require that the triplets $(\mathbf{E}, \mathbf{B}, \mathbf{V})$ associated with matter are left-handed. Correspondingly, the required change of sign has to be reported in the model equations. Let me recall again that the spinning of a vortex ring is definitely different from that of a body revolving around an axis. The more sophisticated ring interpretation offers a clear way to redefine the concept of spin in line with observations[66]. The built-in asymmetry of electrons and protons influences the shape of atoms and molecules. It is evident that the mere analysis of the electrostatic forces does not give a clear picture of the organization of a complex molecular structure. It is at least necessary to introduce considerations regarding the displacement of the magnetic fields, which are responsible for orientation angles and symmetry breaking. I will deal again with this problem in the coming chapter.

So far we have insisted on the fact that an electron must be a positively charged particle. Let us see now why. The orientation of the fields influences three factors: the polarity of the particle, the direction of the magnetic

---

[65]From [Maruani(1988)], v.4, p.73: "In the hierarchically organized levels of life, asymmetry appears at the scale of the nucleus, the small molecule, the chain, forms (organelles, cells, organs, individuals) and stereochemical composition (from the macromolecule to the biosphere)".

[66]From [Preston(1962)], p.385: "Speaking anthropomorphically, one might say that the neutrino can tell a left-handed screw from a right-handed screw. This might not be what one would intuitively expect of such an apparently basic and therefore supposedly simple particle, but it is the experimental fact. Hence there *is* something in the atomic domain capable of distinguish handedness, and it *is* a matter of physical significance which type of axes we choose to describe physical systems interacting with neutrinos".

momentum and that of the rotation axis. If we want the electron to be
negative, we have to impose $\rho < 0$, so that there are sinks in the electric
displacement inside the particle. According to classical physics laws, a
spinning charge produces a magnetic field (see figure 2.5). If we want the
orientation of the electron magnetic field to be compatible with the standard
choice, one has to set the triplet $(\mathbf{E}, \mathbf{B}, \mathbf{V})$ to be left-handed. Therefore,
we find ourselves with two possible settings. The electron is negative, the
triplets must be left-handed, consequently the modeling equations have
to be corrected to match the change of sign of the electric field. As a
second option we have that the anti-electron is negative, so that the triplets
must stay right-handed and we can keep the equations as we are used to.
Inevitably, the last case implies that the electron ends up to be positive.
When the electron was discovered, the choice of its polarity was totally
arbitrary. Now we observe that this choice does not agree very well with
that of the cross product based on the right hand[67]. In the end, it is just
a matter of notations, but if we do not want to subvert some elementary
convention rules of differential calculus, it would be better to assume that
the electron is positive (and the positron negative, I beg your pardon for
the oxymoron).

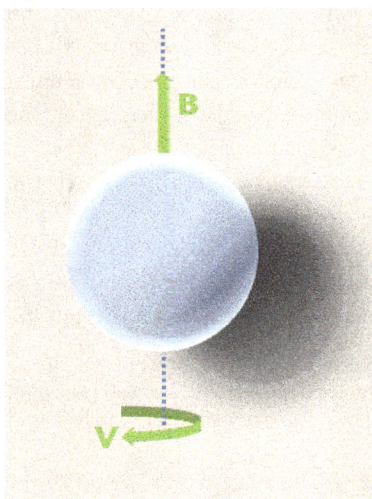

Fig. 2.5 *A negatively charged rotating dielectric produces a magnetic field oriented as shown. Having $\rho < 0$, the electric field points
toward the center of the body, so
that the triplet $(\mathbf{E}, \mathbf{B}, \mathbf{V})$ turns out
to be left-handed. In order to get
a right-handed triplet, the charge of
the object must be positive.*

---

[67] A Murphy's law says: "You can never tell which way the train went by looking at the
track".

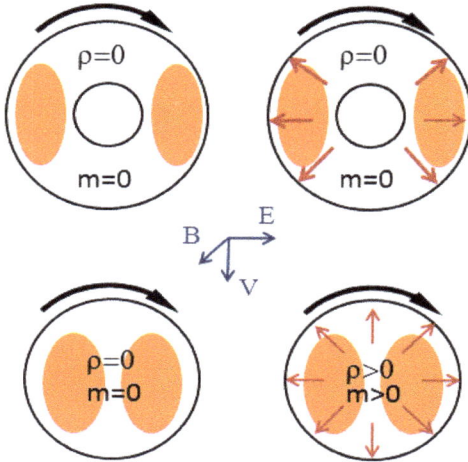

Fig. 2.6 *Schematic representation of the sections of three different types of particles. We can have a single rotating photon in a hollow toroid, with the possible addition of a stationary electric field with $\rho = 0$ (first two pictures). Otherwise, we can have a rotating photon in a solid toroid (third picture). Finally, the last picture refers to the section of an electron.*

These disquisitions on antimatter allow me to introduce other representatives of the subatomic world. It is useful to code the various situations we shall encounter with the help of a schematic approach. The examination of the section of the toroid already contains enough information to draw the main conclusions, therefore a qualitative sketch of the principal features of a particle can be obtained by the simple pictures shown in figures 2.6, 2.7 and 2.8. In the first one, we can see the ring sections of some kinds of neutrinos and an electron. The most basic neutrino is composed of a single rotating photon ($k = 1$ in the cylindrical solutions of appendix G), though other different circumstances may occur (see [Funaro(2014)]). Its sinusoid aspect is graphically shown by two bumps (see figure 4.4, left). It is a Maxwellian wave, so that $\rho = 0$. We also know that the mass m is zero. There is a direction of rotation indicated by the arrow, although there is no need to introduce the vector field **V**. Stationary electric field may be added; in particular when the hole is present one can impose $\rho = 0$. The neutrino with no hole is the backbone of the electron; it also has $m = 0$ and $\rho = 0$. Finally, the bare electron has additional properties. It displays a constant electric density $\rho > 0$ and a positive mass. The arrows inside the circle give an idea of the distribution of the electric field. The velocity **V** is now rigorously defined, as well as a magnetic field orthogonal to the page, in such a way to form right-handed triplets (**E**, **B**, **V**). In reality, **E** and **B** do not remain firm but, being added to a time-evolving solution, they oscillate in the neighborhood of a steady configuration. It is natural to assign a spin equal to $\frac{1}{2}$ to all these objects, though, without any quantitative consideration, this choice just turns out to be a convention.

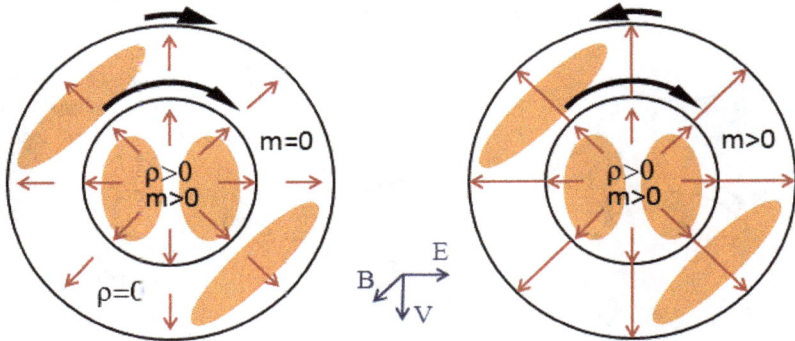

Fig. 2.7   *The bare electron of figure 2.3 is encircled by a secondary vortex ring. In the first case, the outside charge density is zero. In the second case the outer vortex is counter-rotating and its charge density is not zero. This corresponds to the creation of further mass.*

As far as figure 2.7 is concerned, we see in the first picture a bare electron surrounded by a secondary slower vortex with radial electric field having $\rho = 0$. The drawings are qualitative, so that the effective shape of the section of such a secondary vortex may not be a circle. Scales and velocities are also not respected. The left picture of figure 2.7 proposes the same combination of figure 2.2. The formation of an outer vortex looks like an initial attempt to transfer the charge from the bare electron to the outside world. Note that this can be only done if energy is present in the surroundings. Of course, in addition to the two already available, one can assume the existence of further encapsulating rings. Thus, the process of transferring the charge can be achieved by quantized stages. Anyway, this is a problem that will be faced in chapter three. Nevertheless, there is another way to add stationary solutions to the secondary vortex (second picture of figure 2.7), ensuring the continuity of the stationary fields **E** and some (unknown at the moment) compatibility condition for **B**. Note that the circulation of the external layer is opposite to the inner one. Therefore, the triplet $(\mathbf{E}, \mathbf{B}, \mathbf{V})$ displays a different chirality. This can be done by taking $\tilde{\rho} = \mathrm{div}\mathbf{E} \neq 0$, implying the onset of new pressure $\tilde{p}$. In this way, the original electron ring turns out to be covered by a sort of charged and massive envelope constituted by antimatter, although the electric field at the boundary of this object is pointing outward. Thus, the charge looks negative in a left-handed space. The final result is a two-layered bare elementary particle displaying the same charge as the electron. Note that there are now two independent sets of photons and the proportion of their frequencies can be determined when searching for some equilibrium configuration. It would be nice to check instead that such an equilibrium does not actually exist, or, if some balance can be theoretically achieved, it is very precarious. In fact, my guess is that the new particle is very unstable and decays in an electron and an anti-neutrino. The last one escapes like a photon at velocity c, behaving as a free-wave. In the renewed configuration the integral of $\tilde{\rho}$ on the support of the photon turns out to be equal to zero. This decay behavior is known in particle physics and the particle represented in figure is the negative *pion* π^- (π^+ if the interior is constituted by a positron). A possible reaction is: $\pi^- \to e + \bar{\nu}_e$, where the bar on the symbol of the neutrino indicates antimatter. This event is extremely rare. We can see later (figure 2.8) that things may end up to be more complicated.

It is clear that the same principle can be repeated, by adding a third layer, this time composed of matter instead of antimatter. Again, we may obtain a unique particle of charge e, that behaves externally exactly as an electron made of matter does. We can play with the global internal pressure by imposing that its gradient is zero at the surface of the entire body. The inner boundary of the third vortex exerts a force on the inner ones by contracting them with respect to both the minor and the major diameters. We can then assume that a sort of equilibrium is reached, so that the whole system should be able to find an appropriate balance, stabilizing the shape of the separation and the external surfaces. In addition, one can require that density of charge is adjusted in such a way that the total charge is that of the electron (without modifying the constant μ in the model equations). This is reasonably possible if we recall that the entire annular object is a double connected region excavated in the 3D space, and in section 2.1 we established a sort of connection between this topology and the invariance of the charge e.

This alternative version of the electron might be the famous *muon*. This is an *unstable* particle (with a relatively long life however) that shares all its properties with the electron, except that it is more massive (about 206 times more)[68]. Instability is a vague concept in particle physics. Indeed, it is not possible to distinguish among a truly unstable object (that exists mathematically, but unable to stay in equilibrium independently of the applied perturbation) and an effectively stable object with a very small stability basin. This is a kind of analysis that could be rigorously done on my muon model, but with an extreme waste of computational resources.

The study of the muon mass can be done more easily; it is the question of integrating ρ_m in the various domains. There is anyway a difficulty. The function p may be uniquely determined by setting the pressure to be equal to zero at the center of the inner toroid. Instead, the pressure inside the two other vortices is determined up to additive constants and I have no reason to conclude that the global pressure has to be continuous across the common interface (realistically, this should not be the case). To say the truth, I expect some jumps of p at the separation surfaces. Indeed, the difference in velocity of the internal and the external photons produces a discontinuity in the curvature properties of the 4D space-time when approaching either side. We also said that the scalar curvature R must be proportional to p. Hence, more quantitative results can be given only after a serious pondering of what actually happens at the interfaces. The lack of information about the behavior at the boundaries should not prevent us however from advancing conjectures and continuing the discussion. Anticipating some of the results of section 2.6, the essence of my reasoning can be summarized as follows. The discontinuities of pressure witness that something important happens on the separating surfaces. Energy is concentrated there, that in the classical sense can be interpreted as something due to surface tension. Breaking a surface can liberate this energy, with a pop that recalls that of blowing soap bubbles. The process alters the global topology and, at the same time, modifies masses.

Hitting a muon in the proper way can destroy its outer rings with a drastic reduction of mass. The energy will always be preserved; it is just going elsewhere, perhaps momentarily stored in some neutrinos because of momentum conservation. The most significant remnant of this interaction

---

[68]From [De Benedetti(1964)]: "... but the muon is a puzzle: nobody knows why its mass should be different from that of the electron, since muons and electrons are identical in all other ways. Furthermore, at present state of our knowledge one has the feeling that the world would be essentially the same if the muon did not exist, and thus this particle remains both unexplained and unneeded". I agree on the statement that the muon is unnecessary; it is however important to have it in order to comprehend how mass is assigned to particles. If it was not yet discovered, my approach would be able to theoretically predict its existence.

is a muon without its cover, i.e., an electron. Such a subatomic process corresponds to that effectively documented by laboratory experiments. A complicated recombination, schematically shown in figure 2.8, may transform a pion into an electron in a few steps. From my viewpoint, the final products are then constituted by pure electromagnetic components (including the electron). The possibility that photons are the union of a neutrino plus an anti-neutrino is examined for instance in [Raychaudhuri(1986)]. The whole process may be regarded as an evolution of 3D rings, moving according to the laws of fluid dynamics. It is not easy to figure out all the steps in detail, but I will try to provide further explanations at the end of this chapter. Nature offers however numerous suggestions to interpret these phenomena at various scales of complexity[69], up to the fantastic elaborateness of *turbulence*. The two neutrinos ν_e and ν_μ of figure 2.8 are purely indicative. Their exact organization largely depends on the causes responsible for the breaking of the muon's sealing envelope. Everything also depends on the environment. Going ahead with my exposition, it will always be more evident that particles are far from being "naked"; they are dressed by an ocean of electromagnetic signals, like a tornado in the atmosphere[70]. Certainly, the rules of mechanics must hold, and momentum conservation belongs to this category. As far as spin is concerned, the question is a bit delicate. Based on what decided in section 2.1, I can define a spin axis that does not correspond to the usual notion, but that allows me to distinguish between "up" and "down". By looking at the toroid sections, I suggest to assign a positive sign if the object (made of matter or antimatter) rotates clockwise. The sign is going to be negative when the rotation is anti-clockwise. Let me add also the convention that at the meeting of matter and antimatter the corresponding peripheral velocity vectors **V** must assume opposite directions. According to the above definitions and in agreement with the standard assumptions, pions have spin equal zero and muons have spin equal $\frac{1}{2}$. In the decay process of figure 2.8, nuclear physics suggests the existence of a particle W (not well identified in my case).

---

[69] From the introduction in [Shikhmurzaev(2008)]: "It is in the flows with topological transitions where the free surface manifests itself in the most spectacular way, making both experimental studies and the mathematical description of fluid motion so notoriously difficult. At the same time, due to the complex nature of these flows there are still many amazing effects waiting to be discovered and understood even in seemingly simple phenomena we see every day".

[70] From [Fritzsch(1983)], p.148: "This is what physicists call a physical electron: an electron and its vacuum polarization cloud. An electron without a vacuum polarization cloud is called a naked electron".

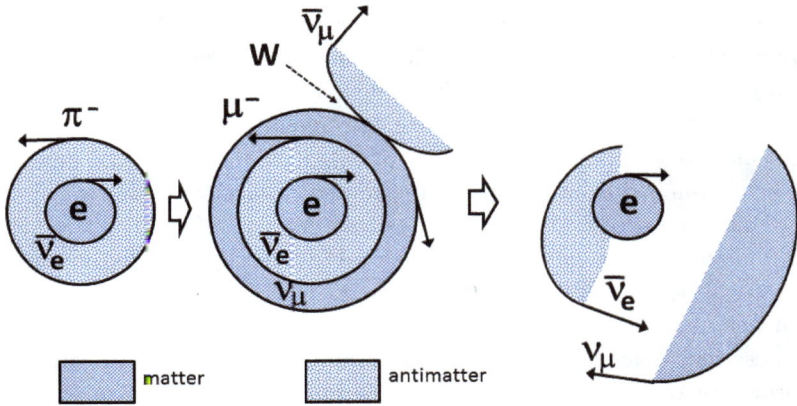

Fig. 2.8 *Particles typically involved during pion decay. Due to the pairwise cre-*
ation of a neutrino ν_μ and anti-neutrino $\bar\nu_\mu$, we have the formation of a negative
muon μ^-, which is a more stable particle. This transition is "mediated" by a
very ephemeral particle called W boson. The muon however does not last for
long and decays into its components. The external shells break up, destroying
the topological environment. As a consequence, the extra mass of the muon dis-
appears. At the end of the process the electron remains isolated. Formally, the
reaction takes the form: $\pi^- \rightarrow \mu^- + \bar\nu_\mu \rightarrow e + \bar\nu_e + \nu_\mu + \bar\nu_\mu$. In other circum-
stances, we can have the direct decay: $\pi^- \rightarrow e + \bar\nu_e$. The whole recombination
phenomenon recalls that pertaining to the formation of vortices and anti-vortices
in fluids.

There is a very short transition period, during which the electromagnetic information stored in the muon's secondary vortices is reorganized.
Preserving a global continuity, the electric and magnetic fields mutate their
displacement, by stretching and twisting. Pressure diminishes until it is different from zero only in the final massive body left: the electron. Communications between the various parts are made possible via the electromagnetic
background. Note that these kinds of transitions occur systematically at the
atomic level and they constitute the environment for complex interactions
at a larger scale. Thus, they should deserve a lot of attention. Outside the
muon, further shells can be built. They can be "inert", meaning that they
carry fields having zero divergence, able to transport the muon's charge far
away. They can be "active", i.e., contributing with other massive slices.
The short living *tauon*, the heavier companion of the muon, might come
from this last process.

Under the name "neutrino" I collected infinite types of objects that I used (and I will continue to use later) as starting skeletons for more complex structures. Included in this family are the neutrinos actually detectable in high-energy subatomic reactions. These do not seem to respond to electromagnetic perturbations and, in the form of free-waves, are able to travel inside matter without apparent resistance. Their behavior is different from that exhibited by classical photons emitted by excited atoms. So, what is the real discrimination between a plain photon and a photon-neutrino? Unfortunately I have no answers. Perhaps, neutrinos are much smaller in size than their brothers. In the direction of motion they are modulated by an entire period of sinusoid, instead of displaying a single peak. Finding solution to this question is crucial, but at the moment I have no convincing explanations.

2.4 Protons

The next and more ambitious step is to build the proton. Here the computations get far more complicated and let me honestly admit that I do not have a rigorous proof of all the claims. I will conduct a study on the feasibility of the project of creating proton-like particles, arguing with classical tools as have been done to this moment. The conclusions emerging from high-energy laboratory experiments indicate that proton structure is rather complex. Therefore, the model has to reflect such an intricate behavior. The conjectures here advanced could be tested with a series of numerical computations. This achievement, although not immediate, is not too hard. The setting up of an effective numerical code requires however time and professional skill.

The new particle is also expected to have a toroid shape. In this respect it will be isomorphic to the electron, carrying in absolute value the same charge (recall the topological interpretation given in section 2.1). But this will be the only similarity. The difficulty in the new construction is that the ring is going to be quite deformed to the point that its hole can be reduced to a segment. In fluid dynamics, structures of this type are known as Hill's spherical vortices. If in the case of the electron the section is almost a circle, the determination of the section's shape is a demanding exercise in the new type of vortex. I also cannot exclude that the section is not of steady type, but freely oscillates around an intermediate configuration with periodic motion. The most significant part is however the interior and, regarding this, I am going to follow a track suggested by some proven facts plus a dose of intuition, although there is also room to formulate alternative hypotheses.

It might sound weird, but the base for operations is going to be again the bare electron. The first idea is to change the polarity of the electric field, so switching the sign of ρ. However, this strategy does not lead to any stable configuration. More precisely, when dealing with antimatter (see the previous section), such an option will be allowed, reproducing in this way the *positron* (the antimatter conjugate of the electron). Remaining instead in the framework studied so far, the introduction of a stationary component having electric and magnetic fields of different signature with respect to the electron (so replacing γ_0 with $-\gamma_0$ in (108)), raises some problems. Going through the computations, one realizes that it is impossible to impose that the normal derivative of the pressure at the particle's boundary is zero as done in appendix H (see (113), implying that γ_0 must be negative, since $\delta > 1$). Following this idea, one finds that there is no balance between the centrifugal acceleration and the electrodynamical component given by the expression $\mu(\mathbf{E} + \mathbf{V} \times \mathbf{B})$ (see (87) in appendix G). As a consequence, the body tends to inflate, squeezing the central hole. Even assuming a bound, due to self-interaction and the consequent development of inner pressure, a stable state seems however not achievable. This is in contrast with the naive prediction I proposed in [Funaro(2008)], section 5.4. In absence of a more thorough analysis, that proposal, perfectly fitting the electron requirements, was still too qualitative for the proton. In the end, it is a good discovery that theoretically the proton fails to be so simple. In some way, this is an indirect confirmation of the model: if the trivial proton existed, the theory would disagree with experiments.

In order to build a proton, there is the need for a robust scaffolding. For this reason, we insist one more time on the stable bare electron. Provided a suitable amount of external energy is present, we assume that photons are dragged in rotatory motion in order to wrap the electron forming a Hill's vortex presenting a kind of ellipsoidal shape. The problem of determining Maxwellian waves of this type has been numerically solved in [Chinosi, Della Croce and Funaro(2010)], by looking for suitable eigenfunctions associated with the vector wave equation. This exercise has been explicitly solved in a complex geometry, constituted by the excavated Hill's vortex deprived of the internal electron ring. For the reasons there explained, the dynamical behavior is obtained from a time-dependent linear combination of two eigenfunctions, corresponding respectively to the fourth and the fifth eigenvalues of a suitable elliptic differential operator. The computational domain is such that the above eigenvalues are equal. This means that the central torus must have a prescribed diameter and occupy a fixed position inside the secondary torus. These configurations are very rare; therefore, the difficulty of the computation is detecting such peculiar situations. The procedure can be taken as a general recipe to find exceptionally weird solutions of the entire set of Maxwell's equations, generalizing the simplest one defined on the most basic toroid.

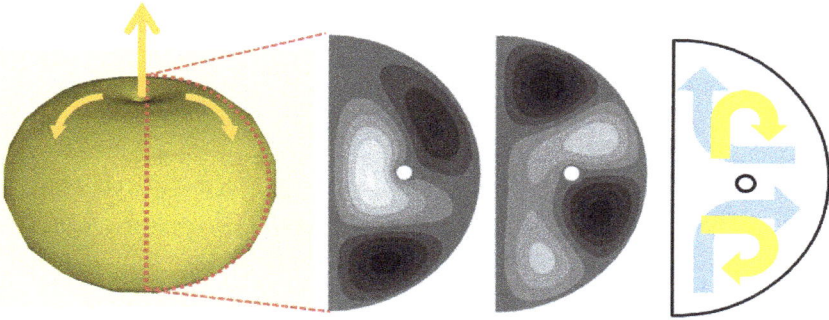

Fig. 2.9 *Maxwellian periodic waves can be found in complicated domains as the above one, where a secondary fat toroid encircles a more standard one (determined by the small hole shown in the three sections). Two steps of the evolution are plot, showing the intensity of the electric field. From these, one recognizes the correlated rotatory movement of two photons (each one represented by black and white spots). The arrows qualitatively indicate the phases of the evolution. The photons are not completely untied but present time-dependent overlapping areas.*

In the specific case we are dealing with, the solution consists of two almost independent spinning photons exhibiting a phase difference of 90 degrees, interlacing in a complicated dance (the animations that can be found in my *web* pages are more descriptive than a thousands words). A rough idea of what happens is given in figure 2.9. When the shape of the secondary toroid is a perfect sphere, from the experiments one recovers that the frequency associated with this periodic evolution is approximately equal to $c\sqrt{136.3}/4\pi d \approx .93c/d$, where d is the diameter of the outer ring. In the end, the structure will basically contain two distinct sets of circulating photons: one inside the internal ring and one floating outside. Using a term borrowed from fluid dynamics, the secondary lower-frequency photons correspond to "recirculation" around the primary fast rotating vortex. As in the case of the single ring, at the outer boundary, the magnetic field is zero and the electric field is lined up with the direction of motion (i.e., it follows the meridians).

Examining the numerical solutions obtained in [Chinosi, Della Croce and Funaro(2010)] by solving Maxwell's equations in toroid-shaped domains, one finds a correlation between the thickness of the central annulus and that of the secondary ellipsoidal toroid. When the minor diameter of the internal ring tends to zero (infinite frequency of rotation) the vertical axis of the ellipsoid ends up to be .60 times the horizontal width. The resemblance with experiments on fluids is astonishing. Infinite other solutions might be studied by assuming that the proton's shape is not exactly an ellipsoid. Nevertheless, one has to consider that Maxwell's equations introduce heavy constraints, so that there is a strict relation between the form of the inner and the outer surface. The mathematical analysis of this nice and difficult problem is a stimulating *shape-optimization* exercise, independently from the possibility of applications. Let me remind the reader once again that these are not speculations, but effective solutions of the Maxwell's system. Let me now enter the conjectural part. In fact, the next and final step is to dress such a dynamical structure with a component displaying $\rho \neq 0$. Hence, the general set of model equations has to be taken into account. In this regard, I have no numerical confirmations about the feasibility of this operation. Nevertheless, a description can be attempted.

Adding a part with $\rho \neq 0$ amounts to providing the entire object with charge, magnetic properties and a brand new mass (see previous section). Let us suppose that the inner ring behaves as a stable electron. In this circumstance, we have a radial diverging electric field with ρ constant and greater than zero (remember that I said that an electron is positive). We then prolong the field outside with continuity, whereas the velocity field will suffer a discontinuous gap. The purpose is to argue as in the case of the muon, so generating a charged and massive surrounding of the bare electron. Pressure is the main ingredient for creation of mass and, at the same time, pressure will also act on the inside electron by modifying its size, until an equilibrium is reached. On the farther outside boundary we need instead to require the normal gradient of pressure to be zero. It is reasonable to assume that the fields are not uniformly distributed on the surface. In addition, both the electric and magnetic fields are expected to be zero along the main symmetry axis. A schematic representation of the section of a proton (negative charge in a right-handed space) containing a revolving inner ring (not to scale) comparable to an electron plus two recirculating photons is visible at the bottom of figure 2.10. The magnetic field is orthogonal to the page in such a way that the triplet $(\mathbf{E}, \mathbf{B}, \mathbf{V})$ is right-handed. Note that $\mathbf{B}$ always points in the same direction, as $\mathbf{E}$ and $\mathbf{V}$ are switched together. Nevertheless, I do not expect that the solution we are looking for is trivially the sum of independent dynamical and stationary parts. Indeed, the problem is hard and chances for simplifications are remote.

An important theoretical result (and a significant validation of my conjectures) would be to show that, given the constant μ as in (89) (appendix G) and assuming the total charge to be equal to $-e$, the global mass of the newborn object actually agrees with that of the proton. The difficulties inherent with these computations are similar to those described in the case of the muon (see previous section). Another achievement would be to discover that the diameter agrees with the experimentally estimated quantity: 1.8×10^{-15} (1.8 fm). Of course, if these facts were true I would be incredibly happy.

As I said, photons trapped in unconventional vortices may assume interesting dynamical behaviors, where one does not necessarily recognize an effective rotation around a common center. Moreover, by adjusting the shape of the ring, the internal evolution may be altered so that the photons follow different patterns. On the other hand, it is well known that the proton is a very complex particle, displaying nontrivial substructures. This is one of the main issues that led to the introduction of *quarks*. Differently from the electron, which is extremely flexible and resistant to deformations, and seems not to have peculiar internal organization, the proton looks more rigid and compact but with a complicated internal animation. It shows however some elastic properties and, when not at rest, assumes multiple forms[71].

Most of the known properties about protons follow from high-energy experiments with large colliders. The *deep inelastic scattering* of protons reveals recurrent behaviors, suggesting the existence of three primary constituents (two *up* types quarks u and one *down* type quark d), members of a family from which matter is supposed to be generated. Quarks u and d have a charge equal to $-\frac{2}{3}$ and $\frac{1}{3}$ of the electron's charge, respectively. The three sub-entities cannot be isolated from the context they live in. There is an exchange and recombination of quarks in nuclear reactions, in

---

[71]From [Miller(2008)]: "For high momentum quarks with spin parallel to that of the proton, the shape resembles that of a peanut, but for quarks with anti-parallel spin the shape is that of a bagel".

a repetitive way that gives strength to the theory of the *Standard Model* and suggests a certain number of basic interaction rules (see [Hoddeson et al.(1997)] for a review). Recent electron-proton scattering experiments also reveal internal magnetic properties of the protons, attributed to the movement of the charged quarks. There is however no agreement between the spin of a proton a that of its constituents. This problem is known as *proton spin crisis*, (see, e.g.: [Jaffe(1995)]).

At this point of the exposition, it should be clear to the reader what is my opinion about particle structure; in particular, proton internal organization does not comply with the existence of subparticles. As far as I am concerned, the proton is made of a bunch of trapped photons, spinning approximately at the speed of light in a very tiny region of space and pro-

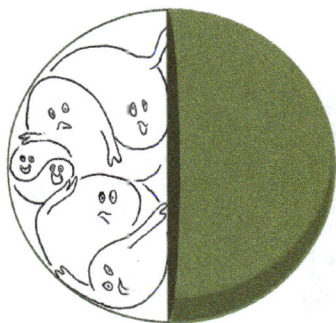

viding the particle with charge, magnetic momentum and mass, in a way similar to that already described for the electron. The special geometrical setting, makes the recipe of the proton more complicated than that of the electron, but the basic ingredients are the same. Thus, the most elementary brick remains the photon, carrying electromagnetic information and moving like a fluid.

I do not deny that experimental evidence suggests, through the detection of peaks in the statistical response, the possible existence of substructures, showing up in certain resonance conditions. There is no need however to look for a specific individual naming of the various constituents, since they are all expressions of a global unique phenomenon and are automatically destroyed when the mother particle is broken. Charge is also existing globally and cannot split up. Therefore, that quark d carries minus one third of the unitary charge, and the two quarks u carry four thirds of it, does not look very meaningful in this context, although one can effectively recognize from my proton model that the total charge is the sum of positive and negative contributions.

The fact that a group of quarks may split and rejoin in different manners, based on coded algebraic laws, does not testify to their effective presence. Substructures in a complicated flow also undergo similar decomposition and recombination processes. The parts mix up with continuity before reemerging in a modified fashion; in other situations they experience changes of topology through the creation or the breakage of fictitious separation membranes. The real novelty here is that such fluids carry electromagnetic information; thus, together with the standard concepts of density, pressure and vorticity, one can discern electric and magnetic properties. In addition, from my viewpoint, there is no need to introduce *gluons* (other elementary particles, acting as exchange particles for the "strong" force between quarks) and the emission of gamma photons from a nucleus will turn out to be a very natural electromagnetic process. Nevertheless, there is the need to show, with more accurate calculations, that the structure I am proposing here resembles (beyond a reasonable doubt) that of a real proton, and this is still a weak point of my description.

There are thousands of papers and sumptuous budgets in particle physics[72], thus, what I am writing here is certainly heretical to the ears of most of the researchers in the field. The quark model agrees quite well with observations and many predicted results have been lately confirmed. There are however many open questions left[73]; first of all, the quark model looks completely disjointed from reality and does not let us understand the end of the story, since work is still in progress. Then there are philosophical questions. What are quarks made of?[74]. How does the mediation mechanism work, during interactions? How do they combine with electromagnetism? My alternative explanation is not consolidated by a long history of research and I am sure that, at this early stage of development, it can be easily exposed to the attacks of experienced particle physicists. At the moment, I can only reproduce with classical tools some features of the main elementary particles. I consider it an important achievement that this can be done with the help of partial differential equations, a mathematical approach that has been almost forgotten in particle physics. A straightforward check of the validity of the model can be obtained by applying numerical techniques, with the goal of getting approximated solutions to the equations that compare well with the experiments. Unfortunately,

---

[72]From [Nambu(1985)], p.10: "The symbol of elementary particle physics is the giant accelerator. Without this tool there can be no particle physics experiments, and without experiments there can be no progress in physics".

[73]From [Nambu(1985)], p.28: "... the quark is one of the particles thought to be 'elementary' at present, and still has not escaped the realm of fiction completely".

[74]From [Fritzsch(1983)], p.263: "Perhaps there is one more level of structure, and leptons and quarks are composite systems consisting of as yet unknown subunits. Such lepton-quark subunits have already been introduced into the physics literature, bearing such names as subquarks, preons, stratons, rishons, and haplons. [...] There may even more levels in the hierarchy of substructures, nothing is known at present".

even for the most elementary configurations, the computations are rather massive, so that the problem is going to be postponed.

Remaining within the framework of the quark, let us look at results regarding the disintegration of protons. In fact, significant indications are experimentally obtained from the encounter of a proton and its antimatter version (i.e., the *anti-proton*), where a series of short-living conglomerations of subparticles are observed. Similar conclusions may be achieved through my model. An anti-proton contains one quark of type $\overline{d}$ and two quarks of type $\overline{u}$. When particle and anti-particle meet all their components mix up (see figure 2.10, where, in a right-handed space, one has to remember that all charges have sign opposed to the usual ones). The couple $\overline{u}d$ joins to form a negative pion π^- (positive, according to my notations). The couple $\overline{d}u$ joins to form a positive pion π^+ (negative, in truth). Finally, the couple $\overline{u}u$ produces a neutral pion π^0, that, most of the times suddenly decays to a bunch of photons. Let us start by saying something about this last case. Photon production is quite natural in my context, where everything derives from electromagnetic radiation. The disintegration of π^0 confirms in some way a possible affinity between neutrinos and electromagnetism (not confirmed by physical experiments), though, as said at the end of section 2.3, I cannot add at the moment more specific information about it.

Positive and negative pions display very short life and in high percentage decay into muons and neutrinos (see the reaction shown in figure 2.8). Pions may also directly decay into electrons and neutrinos, and this is also compatible with my setting. All these processes involve alterations of masses, which is an uncontrolled procedure within the context of particle physics, where one must include the kinetics. As far as I am concerned, masses are modified because the topologies of the various objects change; so, every time a separation membrane is broken (see section 2.2), sudden variations of masses are noticed. Such a crucial subject is going to be reconsidered in section 2.6. The collision of a proton uud with an anti-proton $\overline{u}\overline{u}\overline{d}$ is explained in figure 2.10 in terms of vortices. The smashing of the two particles leads to the formation of three *mesons*: $\pi^- = d\overline{u}$, $\pi^+ = u\overline{d}$ and $\pi^0 = u\overline{u}$, all having spin equal to zero. Spin balance is obtained by attributing spin $\pm\frac{1}{2}$ according to the directions expressed by the black arrows. As mentioned above, the last particle is basically out of interest, since it dissolves pretty soon into photons and anti-photons.

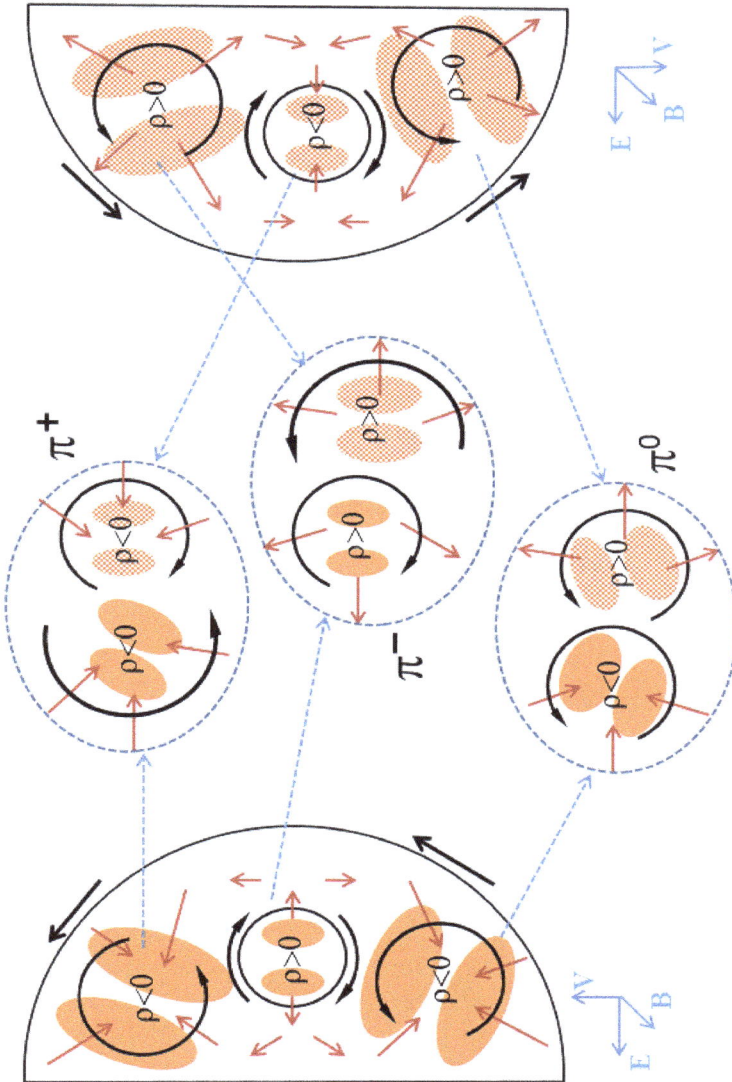

Fig. 2.10 *Fragmentation of a proton (solid) and an anti-proton (dotted). The collision generates three pions: π^+, π^-, π^0, each one consisting of two quarks. The latter is highly unstable and decays into γ-rays. The two others display a short life before decaying in different particles and anti-particles.*

Pions π^- and π^+ are also known for having a short life. I guess that π^- (and similarly π^+) remains for a while the distinct union of the swirls representing the two quarks d and $\overline{u}$. In fluid mechanics, two vortices may adhere and stay attached for a certain time. Usually they start doing a kind of leap-frogging (see section 1.8). After a while, the two structures merge, giving birth to the actual particle. However, the situation for a pion does not last for long. The embryo of a couple made of a neutrino and an anti-neutrino progressively grows for reasons that can be mainly attributed to the state of the environment. One of the two new entities embraces the pion. When the set is closed the intrinsic density of charge, floating within the various pieces, becomes unitary and together with pressure generates effective mass. The resulting particle is the muon (see figure 2.8). The rest of the energy remains stored in the other neutrino. There, the excess of charge does not become mass and it is quickly dissipated in the environment.

Let me provide some other examples. According to figure 2.8, π^- is an electron plus an anti-neutrino. The interaction of π^- and a proton p ends with the creation of a neutron n (see figure 2.11). During this transition, a couple of matter and anti-matter neutrinos are generated. The recombination of the parts involves the formation of a neutral pion π^0, that finally disintegrates: $\pi^- + p \to n + \pi^0$. Thus, it is not necessary to assume that a neutron is of the form udd, as we shall see later in section 2.5. In these transformations momentary mass is created and then destroyed. Indeed, masses are well defined and stable only when they are perfectly sealed in suitable envelopes, while a reaction passes through a series of topological transitions. Any lack or excess of mass is however recorded in neutrinos and free photons in such a way as to preserve energy. The reaction $p + p \to d + \pi^+$, where a model for the deuteron d is given in figure 2.13, may be explained in a similar way (see also $\pi^- + d \to 2n + \gamma$). Similarly to the pion case, neutrinos carrying charge density different from zero may also adhere around heavy particles, such as protons and neutrons, producing even more massive bodies that are actually observed in experiments. This may lead to the introduction of new quark flavors and explain for instance the origin of *hyperons*. A hydrodynamic model of the structure of subatomic particles has been introduced in [Dayton(2012)]. Note that neutrinos are not obliged to remain in the neighborhood of an impact site; on the contrary, they take the form of free photons and travel at speed c (see the end of the previous section). Later, they may interact with measuring instruments, showing their presence through the energy they carry on.

The secret of the parity of matter is another piece of the puzzle that has been set into place and seems to be well handled by my equations, establishing an explicit link between charge conjugation (C) and parity (P), both obtainable by replacing $\mathbf{E}$ by $-\mathbf{E}$. We can include also time (T) reversal in this analysis, since a switch of the arrow of time influences the orientation of the field $\mathbf{V}$. The so called *CPT-invariance* is an experimentally established law that has found large consensus[75]. I believe that my constructive approach may contribute to its explanation. Recall that, in my model, changing parity is not exactly to perform a mirror image, since, as I said above, the positron is not the reflection of the electron. Nevertheless, it is true that positrons are made of anti-photons that are mirror images of photons. There are however experiments that do not appear to be *CP-invariant*. Seemingly, two types of almost identical neutral *kaons*, K^0 and $\bar{K}^0$ exist (see [Segrè(1977)], chapter 15, or [Martin and Shaw(2008)], section 10.2). These are unstable particles with equal mass, displaying two different sets of decay products with different parity (more insight will be given in the next section). The two decay processes do not have the same probability of occurring and this is in strict relation with the velocities of the kaons. Usually, the discussion proceeds with subtle and groundless questions about the reversibility of time and I prefer not to speculate in this direction. Different decaying behaviors of the same unstable particle may be justified in terms of the "environmental influence". There is no way to state universal laws without including the influence, more or less marked, of such an incredible source of energy[76,77]. This may be especially true when the disturbance comes from a recording apparatus, no matter

---

[75]From [Burcham and Jobes(1997)], p.273: "There are no fundamental reasons why the forces in nature should be invariant under the transformations C, P and T separately, but taken together the combined operation of time reversal, space inversion and charge conjugation appears to be a fundamental symmetry transformation which has important and very general consequences".

[76]From [Cantore(1969)], p.188: "When we have an electron moving in the neighborhood of a nucleus, calculations show that it is not enough to take into account the static Coulomb field of the nucleus in order to obtain the energy eigenvalues of the electron, but a coupling with the zero-point vibrations of the field must be also considered. This is what occurs in the Lamb shift, detectable in the spectra of hydrogen and deuterium. Obviously, then, the ground state of the background electromagnetic field plays an observable role".

[77]From [de la Peña and Cetto(1966)], p.194: "This seems to suggest that in general, the accelerated detectors are *not* robust systems, contrary to what is normally assumed in quantum theory: when they are immersed in a local zeropoint field they are distorted by it".

how sophisticated it is. Note that experiments in particle accelerators are supposed to be carried out in "vacuum", a term that needs some clarifications. For example, the *Quantum Chromodynamic Vacuum* (QCD) has been introduced with the aim of justifying a series of pathological situations occurring in high-energy particle physics. This means that the chances to collide with spurious nuclei are very low, whereas the interaction with the invisible background is unavoidable even by taking strict precautionary measures[78]. During their short life, kaons are not singular entities in the middle of a desert vacuum extending between particles. As I will support in the following, kaons are fully immersed with continuous contact in an electromagnetic ocean.

The generation of electron-positron pairs from gamma rays[79], and similar events several times mentioned above, recall the formation of vortices and anti-vortices in the context of fluid dynamics. Such "virtual" particles play a role in the recombination of subnuclear components after collisions[80]. Although made of matter and antimatter, these vortices do not come from nothing. They may originate for instance from the recirculation of some electromagnetic fluid (part of the background I was referring to before) around massive particles moving at high speed, as happens in common flows past an obstacle. Of course, these tenuous suppositions do not quantitatively explain the behavior of kaons (ideas about the property of *strangeness* will be given in the next section). They may however inspire a different way of dealing with subatomic problems. Certainly, as it will emerge in chapter three, the presence of an energetic background is decisive in understanding how particles communicate in order to form atoms and molecules.

---

[78]From [Fritzsch(1983)], p.149: "In processes where the energy of the electron is small relative to its mass, we cannot explicitly see the effects of vacuum polarization. This is what happens in most everyday instances. In these cases we are dealing with the 'dressed' physical electron. However, if the energies involved are sufficiently large (in the scattering process for example) we can see the vacuum polarization effects explicitly".

[79]Following the introduction to *vacuum polarization*, in [Burcham and Jobes(1997)], p.240, we can read: "... QED shows us that the electron can appear in many guises; it can spontaneously emit a photon which may materialize into an electron-positron pair, the electron and the positron may emit other photons and so on. The electron is thus surrounded by a fluctuating 'cloud' of virtual electron-positron pairs with 'radius' of the order of the Compton wavelength ...".

[80]From [Fritzsch(1983)], p.186: "The physical vacuum is filled with an infinite number of virtual quark-antiquark pairs simply waiting to manifest themselves in the form of physical particles. This is feasible, however, only if the corresponding energy needed for this manifestation is available".

In modern nuclear physics, the above mentioned processes are interpreted as a complicated recombination of quarks in agreement with certain accepted diagrams. The aim is to give meaning to a plenitude of various unstable objects, characterized by very short lives (often shorter than the "clock time" of an electron), ejected during heavy nuclear collisions. I do not think it would be difficult to reproduce them in terms of fluid dynamics transition phenomena; there are certainly plenty of degrees of freedom to play with (it is enough to have a look at a whirling water stream to get the idea). The experiments with fluids in [Lim and Nickels(1992)] show the frontal collapse of two vortex rings (artificially colored, the first one blue and the second one red). After the crash, 12 smaller equal rings are formed, scattering orthogonally from the impact site. Interestingly, each one of the secondary rings carries both the colors in equal parts. This "chromo conservation" rule may be a source of inspiration for further characterizations. Analogies between optics phenomenology and fluid vortices, both at experimental and theoretical levels, have been pointed out by many research groups. Here, I am pushing these similarities to the smallest scale. The fact that subatomic structures are repetitive and can be coded into families should not be a surprise, since, according to my theory, there are concise and deterministic differential equations ruling these interactions.

2.5   Nuclei

In order to investigate atoms, a commitment that will start from the next chapter, it is necessary to deal first with their nuclei. To this end, I may just provide some general thoughts, since reliable results can only be given after a serious and systematic analysis, at the moment beyond the scope of this paper. In order to proceed, I need to introduce the *neutron*. This is known to be an unstable particle with an averaged age of 300 seconds. Mathematically, it is unclear if the neutron is really unstable; it may have instead a very narrow basin of stability, statistically broken in a finite period of time as a consequence of external factors. The neutron's mass is slightly larger than that of the proton[81]. At a certain distance such a particle may be considered electrically neutral, but scattering experiments show that it has a positive core surrounded by a negative cover. In the decay process, the proton gives rise to a proton, an electron and a neutrino. The official justification for the existence of the neutrino relies on spin and energy

---

[81] From [Fritzsch(1983)], p.49: "Oddly enough, the situation is reversed with nucleons: the neutral particle is heavier than the charged one. Why this is so we still do not know".

conservation arguments. According to the Standard Model, there should be three quarks inside a neutron. They are two *down* quarks d and one *up* quark u, with total charge equal to zero.

Now, the quark version is quite in disagreement with my theory. Indeed, I think it is more credible to assume that a neutron is actually composed of three separated elementary particles: a proton, an electron and a neutrino[82]. Since we now know many things about their properties, we can easily find a natural combination, fitting the neutron's shape. I propose for instance that of figure 2.11 (see also the front cover of the volume [Shikhmurzaev(2008)] for an amazingly similar UFO profile, in the framework of viscous fluids). The idea that nuclei contain electrons dates back to the earlier stages of atomic physics and was dismantled long ago. Modern physicists are in fact quite skeptical about this version[83], safeguarding as much as possible the quark origin of the neutron. Attempts to give rigorous explanations rely on the *uncertainty principle*[84] that forces the nuclear components round an everlasting roundabout. If such components were at rest, they would be fully "detectable", since position and velocity are known. With the addition of more technical insight, such a situation excludes the presence of electrons in nuclei (see [Williams(1991)], p.14). My model aims at being a deterministic one, so that this type of argument should not apply. The connections with quantum theories and my way of interpreting the Heisenberg principle will be developed starting from chapter three. On the other hand there is already something moving fast inside each particle,

---

[82] From [Miller(1984)], p.156: "Although Heisenberg chose to consider the neutron as a fundamental particle, in the succeeding deliberations he equivocated, owing principally to the problem of where the electrons originated in β-decay. And so, when necessary, he invoked arguments based on conservation of energy - for example, when he discussed the stability of certain nuclei against β-decay. But then, at the paper's conclusion, he suggested that for certain processes such as the scattering of light from nuclei, it is useful to assume that the neutron is a composite particle".

[83] From the introduction in [Preston(1962)]: "The same reasoning which shows that there could not be electrons in the nucleus shows that a neutron must not be thought of as a composite particle, say a proton and an electron tightly bound, but that the neutron must be treated as a fundamental particle on the same footing as the proton. Indeed, it is useful to consider the neutron and the proton as simply two different states of a single particle, the nucleon". This reasoning is based on some discrepancies related to the nucleus of Nitrogen to be discussed later on.

[84] From [Williams(1991)], p.64: "The collective motion of nuclei must be very different from those of a drop of liquid. In the latter case the molecular motions keep each molecule localized to a relatively small part of the drop but in nuclei the uncertainty principle does not permit localization of a nucleon to the same extent relative to the size of a nucleus".

with no hope of escape: the photons. They perfectly satisfy the *Heisenberg inequality*; so I have found a harmonious way to accommodate both classical and quantum-like justifications.

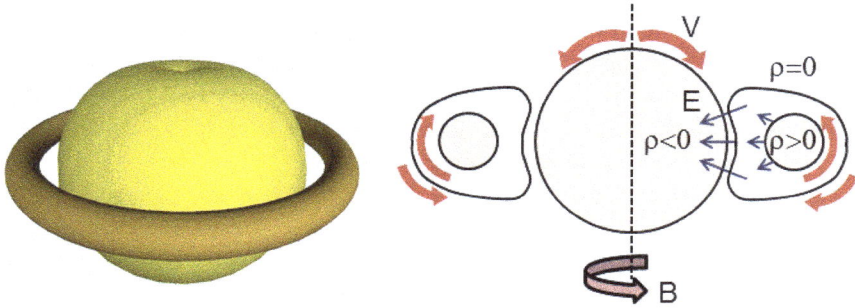

Fig. 2.11   *A combination of a proton and an electron, plus a suitable neutrino, provides the model for the neutron. At a distance, the new particle is electrically neutral, it has about the same mass as the proton and its constituents are exactly those observed in neutron decay. Its global spin turns out to be equal to $\frac{1}{2}$.*

Going back to figure 2.11, the presence of the anti-neutrino $\bar{\nu}$ forms a kind of cushion. The combination of the electron and the anti-neutrino recalls that of the pion π^- (see figure 2.8), though not necessarily one should have: $\bar{\nu} = \bar{\nu}_e$. The negative pion is also present in the reaction: $\pi^- + p \to n + \pi^0$ (where n denotes the neutron), which is correct in terms of quark balance. A way to attribute a spin to the whole particle is by splitting the neutron into its components, the algebraic sum of their spins agrees with the standard one (i.e., the neutron has spin equal to $\frac{1}{2}$). That is also the procedure indirectly followed in experiments, where conclusions are reached after examining the various fragments. Neutrinos are important ingredients both for stabilizing a nuclear ensemble and providing the necessary conservation properties of momenta. They must then be counted for an overall evaluation. For example, the nucleus of Nitrogen ^{14}N, having a real spin equal to zero, according to my view must contain 14 protons and 7 electrons leading instead to an odd parity spin. This rough computation, usually put forth by physicists to show that nuclei cannot encapsulate electrons, does not actually take into account the presence of the additional neutrinos.

Actual measurements agree with the model, since the proton has an experimentally estimated diameter of 1.8 fm (1 fm $= 10^{-15}$ meters) and my electron has a global diameter of 2.7 fm. It has also to be recalled that these elementary components are elastic vortex rings, hence they enjoy a reasonable degree of freedom and adaptability. Moreover, it is documented that the difference between the neutron mass (at rest) and that of the proton is slightly bigger than the electron mass. Recall that the minor diameter of the electron may vary with the inverse of its photon's revolution frequency (see figure 2.1). Touching the proton is however forbidden to the electron; the reason is that the magnetic field at the surface of the proton is order of magnitudes greater than that of the electron. Corrective fields must be added to the intermediate neutrino in order to join continuously those present at the surfaces of the bare particles. The link must be done by respecting the two divergence conditions: $\text{div}\mathbf{E} = 0$ and $\text{div}\mathbf{B} = 0$ (neutrinos are Maxwellian solutions). Hence, the magnetic field has to comply with a sudden variation within the neutrino support and probably this is the reason for the instability of the neutron. Squeezing the neutrino too much may effectively lead to a collapse of the entire structure.

The orientation of the magnetic field inside the neutron ensemble is also indicated in figure 2.11. It is worthwhile at this point spending a few words on the magnetic properties of a particle, since it is an ambiguous issue. Heuristically, magnetic momentum $\vec{\mu}$ originates from the rotatory motion of a charged body, like a tiny sphere revolving around its axis. This classical view is rather in contrast with reality. As a matter of fact, by applying a magnetic field to a standing electron or a proton we do not observe any visible effect. In fact, particles are not magnets and do not display any magnetic axis. Quantum physicists would say that this version of the facts was replaced long ago in favor of a formal combination of mathematical operators, that however take their inspiration from the definition of angular momentum. Thus, it is forbidden to talk about a real rotation, although the theory is build on such a primitive concept (see footnote 106). Modifying experimentally the polarization of a particle seems to be a difficult exercise[85]. Spin detection is usually achieved in an indirect way by examin-

---

[85] From [Tolhoek 1956)]: "One may think of obtaining polarized electrons by a means of a kind of *Stern-Gerlach experiment* sending electrons through a strongly inhomogeneous magnetic field (although still varying on a microscopic scale). However, [...] it is shown that a splitting of an electron beam according to spin orientation cannot be attained

ing the *hyperfine structure* of emission spectra. Other tests determine the torque effect $\vec{\mu} \times \vec{B}$ in the presence of a sophisticated magnetic environment $\vec{B}$. Therefore, what is actually measured is the reaction of the particle to magnetism, which is not necessarily a proof of any spinning-like behavior, or, in milder form, it does not tell us how such a spinning occurs.

According to my interpretation, in the bare particle the magnetic field follows closed lines around the spin axis (coinciding with the symmetry axis of the ring). The situation is similar to that of a current flowing in a toroid wiring. Indeed, it is exceptionally similar, since also at quantitative level the distribution of the magnetic field inside an electron (where computations can be done explicitly) follows the *Biot–Savart law*. By the way, we do not have to be amazed by these coincidences. What is happening in a particle is ruled by the same model that in the large-scale world describes the electric and magnetic properties of matter. The result shows that it is possible to have a rotatory kinetic behavior with no net magnetic flux through a surface orthogonal to the spin axis. When an external magnetic field is applied to my particle, there are actually changes inside its structure, brought through the separation surface. I expect that this stimulation may twist and jerk the particle but are not so serious as to accelerate the whole object by shifting its barycenter, also because, if the perturbing field is homogeneous, the rotatory displacement of the inner magnetic field produces movements with zero average. These are however only suppositions. To know more about this effect, a more careful theoretical analysis is required, also because it would be important to provide a handy formula for evaluating $\vec{\mu}$ that compares well to experiment. It is wise to observe that the magnitude of the magnetic momentum has some sort of relation with particle masses, hence, until an explanation of the meaning of mass is found (see section 2.6), the measured momenta of the various nuclei remain an unrelated set of numbers. I will continue on the subject of magnetic momentum in section 3.1, when discussing atoms. In that circumstance we will learn how important the role of magnetic fields is in stabilizing atomic and molecular structures. Returning to the neutron, the existence of a nonzero magnetic momentum confirms that it is a composite particle (if it was totally neutral one should find no magnetic properties), although in my version quarks only appear inside the proton. Other confirmations in favor

---

in this way: the inhomogeneity of the magnetic field causes a spreading of the charged electron beam (the particles of the Stern-Gerlach experiment are electrically neutral), which is so large that the spreading arising from the different orientations of the magnetic moment in the inhomogeneous magnetic field is not detectable".

of the neutron structure of figure 2.11 come from nuclear reactions, where protons may be transformed into neutrons and vice versa, the difference being the absorption or the release of electrons (or their anti-particles) and high-frequency photons.

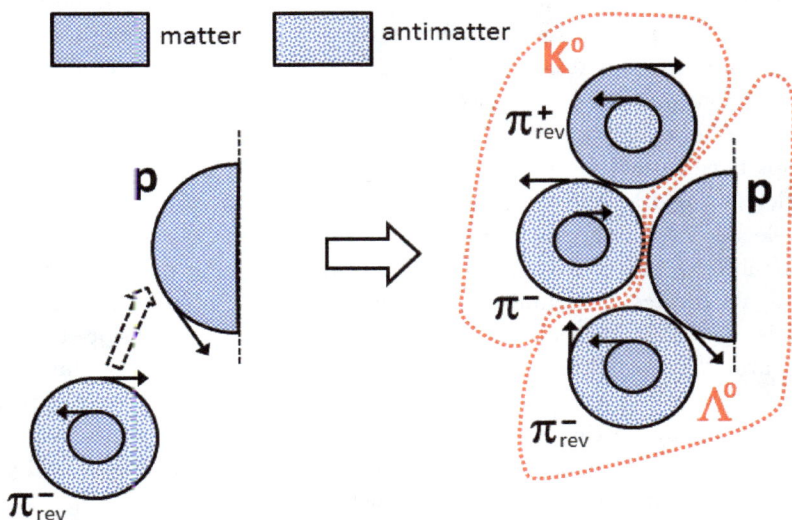

Fig. 2.12   *An encounter of a pion and a proton is not going to generate a neutron as in figure 2.11, if the velocity fields are not properly oriented. The interaction may instead give birth, via the creation of vortices and anti-vortices, to composite particles, that successively decay into their components.*

It is interesting to remark on the following facts. The placement of the arrows in figure 2.11 is made according to a convention established in section 2.4, that at the boundary between matter and antimatter the velocity vectors must be opposite. What happens instead when the orientation of the pion is reversed, so that the tangent velocity vectors at the meeting with the proton are oriented in the same manner? We can, of course, identify the π^- with its upside down version, i.e.: π^-_{rev}. The new particle $\Lambda^0 = \pi^-_{\text{rev}} + p$, having still charge equal to zero and spin $\frac{1}{2}$, is called *hyperon*. When a π^-_{rev} hits a proton, a viable transformation is the following one: $\pi^-_{\text{rev}} + p \rightarrow (\pi^-_{\text{rev}} + p) + (\pi^+_{\text{rev}} + \pi^-) = \Lambda^0 + K^0$, where the couple π^+_{rev},

π^- emerges from the vacuum as a reaction of the incoming unmatched vortex. The final product contains a *kaon* K^0, that has been also mentioned in the previous section. Another possible combination provides the reaction: $\pi^-_{\text{rev}}+p \rightarrow (\pi^-_{\text{rev}}+\pi^-+p)+(\pi^+_{\text{rev}}+\pi^-+\pi^+) \rightarrow (\pi^-_{\text{rev}}+n)+(\pi^+_{\text{rev}}+\pi^-+\pi^+) = \Sigma^- + K^+$ (see, e.g.: [Martin and Shaw(2008)]), where now the neutron has been correctly assembled. Such outcomes are documented by experiments and can be simply coded in terms of circulating and anti-circulating toroid vortices (see figure 2.12, where, similarly to what has been done for figure 2.8, toroidal particles are schematically represented only through an appropriate section). Together with the neutral kaon $K^0 = \pi^+_{\text{rev}} + \pi^-$, it is possible to define its companion: $\bar{K}^0 = \pi^-_{\text{rev}} + \pi^+$, in which matter is replaced by anti-matter and vice versa. Note however that the two versions are not specular, as already noticed in section 2.3 concerning with other particles. The new entities K^0, $K^{\pm}$, Λ^0, Σ^0, $\Sigma^{\pm}$ are characterized by having some contacts between matter and anti-matter where the velocity fields are oriented in the same verse (in contrast with what has been established for more "classical" particles). I am strongly tempted to call this property *strangeness*, because it actually corresponds (same representatives and same conservation rules) to the quantum number S introduced in particle physics[86].

In order to look for further coincidences, it is necessary to proceed with the construction of more involved aggregations. A further step ahead is to build the *deuteron*. This is the nucleus of heavy Hydrogen ^{2}H, composed of a neutron and a proton. There are many ways to proceed. Without the help of numerical investigation, that at this stage of complexity would be an ambitious commitment, I can only propose some ideas. According to figure 2.13, we actually find two protons sharing a single electron. As in more involved nuclei, modifying a bit the position of the electron decides which proton is going to remain a proton or become a neutron. These promiscuous transformations are quite frequent in nuclear reactions and, as far as I am concerned, they are not due to an exchange of quarks, which remain instead confined within each proton in the form *uud*. Concerning the neutron, the quark aggregation *udd* is then definitively abandoned. This is a nice thing after all. In fact, it is not necessary to set the quark charge according to

---

[86] From [Segrè(1977)], p.699: "This whole subject is systematized by introducing a new additive quantum number called *strangeness* (S) (Gell-Mann and Nishijima, 1953) and by postulating that, in strong interactions, strangeness must be conserved. At present, these concepts are unrelated to other ideas in physics, and we have no profound 'explanation' of their meaning, but shall employ them as very useful semiempirical rules".

the proportion $\frac{4}{3}$, $-\frac{1}{3}$. The only restriction is that the sum of the charges of two *up* quarks and one *down* quark must be equal to unity. As far as I know, nobody at present has been able to isolate charges equal to one third of the unitary one.

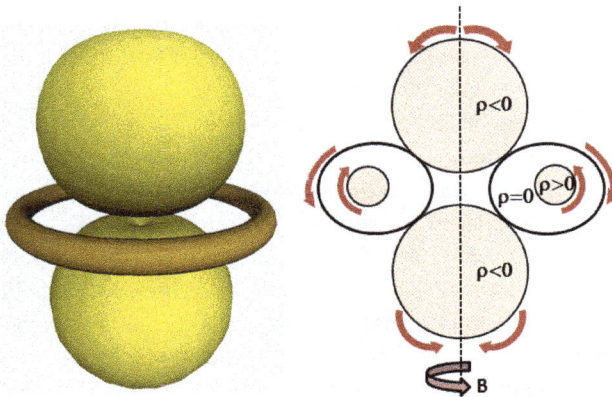

Fig. 2.13 *Possible arrangement of two protons, plus an electron and a neutrino providing the model for the deuteron (spin=1).*

In the deuteron, the various components are held together by a suitable neutrino, produced by trapped photons dragged by the bare particles. The creation of an interstitial pressure helps keep the system stable. Electrically speaking the configuration is admissible, i.e., one can actually place two charged spheres and a ring with opposite charge in (unstable) equilibrium as in figure 2.13. Nevertheless, one should quit the idea that the glue is exclusively based on classical electrostatics. Fluid dynamics is instead the key issue here. This is also the secret for keeping more protons inside a single nucleus. In fact, in my opinion, there is no such concept as *strong force*. In a nucleus, bare protons do not "see" each other, so they are not subject to any repulsive force. The only way they have to communicate is through neighboring electromagnetic signals circulating in between. In small amounts the inner photons help glue everything together; after a suitable threshold they are instead the cause of a vigorous explosion. To know more about the mechanism of repulsion in the case of equal charges the reader is referred to section 3.1. One of the explanations for excluding electrons from nuclei is that their presence, compatibly with other various restrictions, would lead to an extremely high Coulomb energy. Such a concept as Coulomb potential does not apply in my situation, at least not within the range of distances we are examining here. Interaction at a distance between particles are dictated by quantum rules that are going to be studied in chapter three (see also at the end of appendix I). Within a nucleus we are still in a sort of "level zero" regarding quantized stages.

Looking again at figure 2.13, we notice that the various proton surfaces almost adhere, not leaving enough room for undesired intruders. As stated above, each bare particle knows about the existence of its companions only if a source of energy in the form of neutrinos transmits the information present at the various boundaries. If this is done in the proper fashion, the two protons tend to repel but the electron acts as a mediator. However, if further external energy enters the nucleus, it may lacerate the delicate balance holding the structure together. Consequently, a strong reaction is observed. For example, some photons may violate the core by getting stuck inside. The structure "inflates" and the collapse begins, with the result that some components of the nucleus are ejected far away with extreme force. Note that to be effective, the perturbing photons have to be presumably of the same wavelength as the deuteron dimensions (corresponding to frequencies of the order of 10^{23} Hertz). Of course, the destruction process might also be caused by the collision with other particles. The behavior agrees with what is observed in radioactivity and, more in general, in nuclear reactions, where elementary parts, such as electrons (β-decay) or photons (γ-rays) are emitted. In an inverse process, electrons may also be *captured*, becoming part of a nucleus and transforming protons into neutrons. More complex parts, such as α-particles may also be shot out and I will mention them again in a moment. Current explanations of these phenomena are rather involved and require cumbersome technicalities accessible to a limited community of experts in quantum theories. Evidence of competing forces in the proximity of a nucleus are reported[87], and I am more inclined to imagine a non-homogeneous distribution of the charge inside a nucleus, rather than resorting to the existence of inexplicable short-range strong attracting forces, only conceived with the purpose of justifying nuclei compactness. In the end, I think it is not a scandal to suppose that electrons are among the constituents of nuclei; there could not be a better explanation of the β-decay phenomenon! Despite the arguments I give in favor of this

---

[87]From [Eisberg and Resnick(1985)], p.511: "Both the α-particle scattering and the α-particle emission analyses showed that there is a *nuclear force*, which is *attractive*, acting between the particle and the nucleus, in addition to the repulsive Coulomb force acting between the two".

hypothesis and the insufficiency of the arguments of the antagonists; I still have the feeling that many readers are going to repudiate this assumption. Let me go on with the discussion and see if continuing along this path is a feasible alternative or whether one is led to contradictions. In order to validate the scheme of figure 2.13, I wonder if it is possible to effectively generate deuterons from the fusion of two protons plus an intermediate electron. In this experiment the pieces must not be joined together by brutal force. We know that in this case the amount of energy required would be rather large. Nevertheless, if, before collision, one tries to create the conditions that allow the collapsing particles to assume the proper mutual order and orientation of their spins, maybe some results at lower temperatures might be achieved. The trick is to try to follow neutral paths, where, due to the presence of both positive and negative charges, electrostatic effects are momentarily screened. The matching of the magnetic fields must be also realized, since the final magnetic displacement has a crucial role in the stabilization of the final structure.

Concerning again the deuteron, another viable hypothesis is that the two protons, instead of being two separated entities, constitute a single "nucleon" by sharing their internal quarks. The option would roughly correspond to the so called *collective model* of the nucleus in contrast to the *independent particle model*. This possibility brings us back to the nontrivial analysis of constrained photons in regions with multiple holes. The external electron would serve to stabilize the structure, by imposing an appropriate shaping. Indeed, it is well known that the union of only two protons is tremendously unstable. This version better follows the research trend in nuclear theories, although the help of an isolated electron is still required. I am not particularly inclined to this approach, since it does not help justify how charges and masses sum up linearly (for example a structure with 3 protons and 4 neutrons has charge equal to 3 and mass almost equal to 7). From the study of the assemblage of simpler units, the problem now passes to the more demanding computation of the whole unified configuration. One should first detect Maxwellian type solutions providing the structure with a stable support, and then look, compatibly with the model equations, for more general dynamical solutions with $\rho \neq 0$. The reader can imagine how this study is going to be increasingly difficult as more sophisticated nuclei are taken into account. An attempt in this direction must be tried, at least to show that the itinerary is not precluded. The aim of this achievement is twofold. First of all, there is the rediscovery of known properties of nuclei, beyond the algebraic combination of basic elementary components (i.e., quarks and gluons), with the goal of explaining why and how some configurations are stable, possibly providing quantitative results. Secondly, there is the appealing temptation to show that this can be actually done using classical tools (electromagnetism and fluid dynamics).

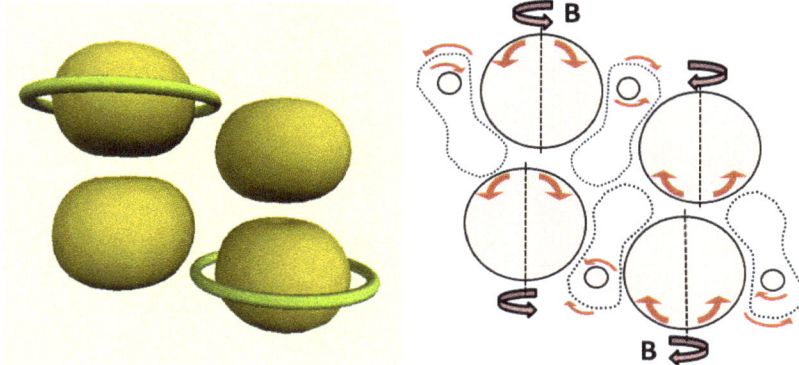

Fig. 2.14 *Hypothetical way to arrange electrons and protons in order to simulate an α-particle. Together with a neat charge of two unities, a stationary magnetic field, displaying zero average distribution, also exists.*

Anyway, for the sake of both simplicity and convenience, I would like to stay on the option of maintaining isolated electrons and protons. So, let me naively combine these main ingredients (together with neutrinos) to form interesting aggregations. From now on the study is purely speculative, although stimulating reflections will emerge. Let me go ahead with the analysis of nuclei by introducing α-particles. These are the nuclei of Helium ^{4}He. They consist of two protons and two neutrons, that, based on my assumptions, make a total of four protons, two electrons and a bunch of neutrinos, suitably encapsulated in a tiny region of space, having an approximate diameter of 4 fm. These composite particles are almost indestructible and are the primary building blocks of most of the more involved nuclei. A possible configuration is the one depicted in figure 2.14. The agglomerate is quite interesting, since it can be naturally and easily combined with other similar geometrical entities.

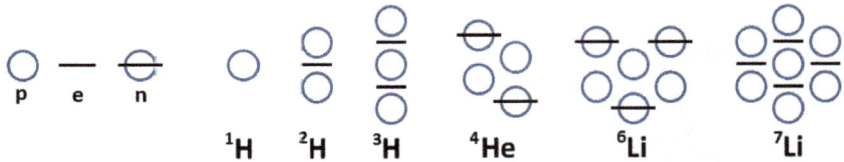

Fig. 2.15 *Schematic representation of some basic nuclei. Protons are denoted by a circle and electrons by a bar.*

The reasons why the structure of the α particle does not collapse from the electric viewpoint are the same as those set forth for the deuteron. Apart from the rushing movement of the constituting photons, the ensemble is not necessarily at rest. In particular, the two electrons may enjoy a certain freedom, being able to move upwards or downwards (in relation to figure 2.14) and assume an inclination. A periodic vibration can possibly shake the whole arrangement, even if it is not under the influence of external factors. I did not define what is the spin of a complex nucleus, but, for symmetry reasons, the global spin of the particle is zero, as it also turns out to be from practical experiments. On the other hand, magnetic fields are present and display anti-symmetric distribution, suggesting that the particle is, to a first approximation, insensitive to magnetic perturbations. Note however, and this is going to be a crucial property in the discussion of atomic structures, that the distinctive magnetic patterns emanating from a nucleus are going to be decisive for characterizing the features of a specific atom (see also figure 2.17). Such a constructive work will be dealt with starting from section 3.1. The way I am proposing to combine the various ingredients of the α-particle remains of course an elegant theoretical conjecture. Nevertheless, this example stimulates further developments. Advancing with circumspection, let me try the assembling of more α-particles and similar gadgets. A preliminary sketch is shown in figure 2.15, where the α-particle scheme is suitably replicated to form Lithium nuclei (neutrinos are not depicted). The displacements of the pieces are not rigorously reported: for example, electrons are free to shift axially in order to find electrostatic equilibrium. The adjustment may require a continuous and systematic readaptation giving the nuclei a restless quivering.

One can easily figure out how the following reactions may occur: $^6\text{Li} + n \rightarrow ^3\text{H} + \alpha$, $^6\text{Li} + ^3\text{H} \rightarrow 2\alpha + n$, $^7\text{Li} + p \rightarrow 2\alpha$. Once again, I recall that, although these configurations might be in equilibrium from the electric and magnetic viewpoints, their stability is given by the electromagnetic energy fluctuating around and passing through the holes of the various rings. The dynamical behavior leads to the appropriate curvature of the space-time. Photons are then trapped in a complex network of chained geodesics and relativistic gravitation is the cement. In this way, we can associate a mass to the nucleus. Nevertheless, if we want to read these geometric interactions in terms of classical masses, suitable clarifications must be made. I discuss this in the coming section.

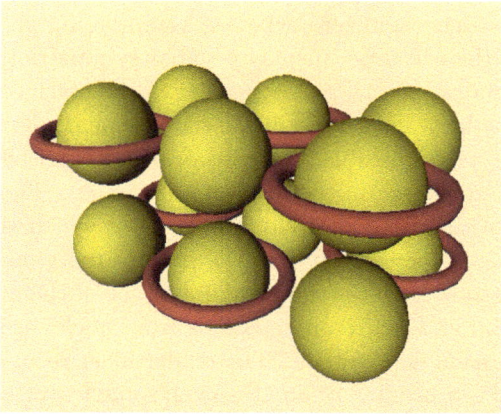

Fig. 2.16 *Hypothetical displacement of protons and electrons in a Carbon nucleus. Neutrinos are not shown.*

More involved situations could be obtained by replicating the ^4He stencil. The example of Carbon is proposed in figure 2.16. The various constituents are closely packed and interlaced to form elegant crystal-like frames. Experiments show that the diameter of the most complex nuclei (including Uranium with 238 protons and neutrons) does not exceed 20 fm (recall that the diameter of a proton is less than 2 fm). The size of the nucleus radius follows approximately the law $1.25 \sqrt[3]{A}$ fm, where A denotes the atomic number (sum of the number of protons and neutrons). Thus, nucleons are sardined optimally. In my view, nucleons occupy fixed positions but photons circulate in and around at speed c, conferring to their union a dynamical behavior compatible with the Heisenberg principle. No strong forces are necessary, since at this level of compactness, repulsion due to Coulomb's law does not take place.

Let us examine other options. The Beryllium isotope ^{8}Be (spin 0) is unstable and decays into two α-particles. To ensure stability an extra neutron must be present (see ^{9}Be). I cannot explain why the modified structure happens to be more solid, but a careful study of different possible displacements of the components may be illuminating. Let me recall that the orientation of the magnetic field must also be taken into account. As already mentioned, the analysis gets more interesting starting from Carbon ^{12}C (see figure 2.16), having spin equal to zero and resulting from the aggregation of three α-particles. Similar considerations may apply to ^{16}O, ^{20}Ne. Using imaginativeness the list can be further prolonged. It should be then clear why the basic isotope of Nitrogen, Fluorine, Sodium, Aluminum, Phosphorus, and Chlorine cannot have spin zero, whereas Magnesium, Silicon, Sulfur, Argon and many of their isotopes display peculiar symmetries. It should also be evident how, under the process of *spallation*, the crash of nuclei can transform an element into an isotope of another element. For instance, by natural radioactivity, the unstable ^{14}C transforms into ^{14}N by losing an electron (and some spare energy in the form of a neutrino). The process involves some redistribution of the participating protons which is not unfamiliar in fluid dynamics, as one can see from the prompt reorganization of an aggregation of soap bubbles after the breaking of one of them. It has to be noticed that neutron and deuteron emissions from nuclei are rarely seen, while α-particle emission is dominant. The reader must recognize that, although the outcome is a product of creativeness, some features of the real world are faithfully reproduced. Fascinating theories assume that heavy nuclei might have been produced from primordial superdense aggregates of neutrons[88]. These packages should not be too dissimilar from the compact clusters here examined.

---

[88] From [Rice and Teller(1949)], p.355: "It is interesting to speculate about such a superdense state of matter and its possible subdivision into nuclei. Here we shall merely state that in this superdense state all electrons are probably pressed into the nuclei, transforming the protons into neutrons, so that the actual aggregate consists of neutrons only. Whenever superdense matter breaks up, beta processes would occur and, together with the emission of electrons, protons would be formed within the nuclei".

Nuclei also display quantum magnetic properties and this does not surprise me, since magnetic fields are automatically built-in and constitute an essential ingredient for stabilization. When I first worked on the problem of assembling nuclei, my primary concern was to quantitatively maintain the amount of spin, according to the specifications given for each element. For instance, ^{7}Li has spin equal to $\frac{3}{2}$. The enterprise was unsuccessful. The spin of a nucleus is not just the algebraic sum of those of the components involved, but also includes the *total angular momentum*, whose definition is borrowed from quantum mechanics. Again, without explicit mention, one comes back to rotating bodies around an axis which is an oversimplified way to study nuclear dynamics, where nonlinearities are predominant. In the end, what really matters is the measured intensity of the magnetic momenta (rigorously and abundantly documented for all nuclei), which do not follow any simple recipe (as it actually should be).

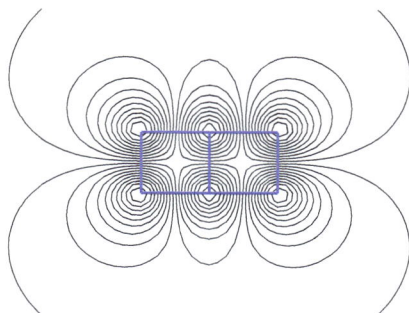

Fig. 2.17
Sketch of the lines of force of the dynamical magnetic field around the Carbon nucleus ^{12}C *of figure 2.16 (top view). The actual displacement is more complex, since the present picture only refers to a 2D projection.*

From a distance, as far as the electric field is concerned, a nucleus looks like a homogeneous spherical charge. The same statement is not correct when examining the magnetic part. The alternate orientation of the spin axes of the constituting particles creates distinguished patterns. It is true that there are no sources or sinks (recall that $\text{div}\mathbf{B} = 0$), but the feeling is different as one moves from one spot to another. This behavior is going to affect the formation of the corresponding atom, with subsequent consequences for molecular bonds. From a trivial massive and isotropically charged body, a nucleus is here promoted to become a fussy controller of the electronic traffic floating around.

Perhaps, the assemblage of nuclei proposed in this section admits improved versions; perhaps, not reliable at all. Independently of the effective construction, the message that I would like to communicate is that all the instructions to compose atoms and molecules are already imprinted in nuclei, that are able to provide the environment with the necessary information both of an electric and magnetic nature. I will examine more deeply in chapter three the consequence of such a revolutionary conjecture.

This discussion is however becoming too speculative. I have too little material (and too little competence) in my favor to review the backbones of nuclear physics, sustaining and adequately justifying alternative approaches. It is enough for me to have shown a possible viable path. On the other hand, the cards of specialists are not always aces; argumentation is based on statistical evidence and if one looks at the way things have developed, there is space for a reinterpretation of the foundations. Inelegant adaptations and exceptions characterize the current theories; so that the discipline, even if introduced in an informal fashion, ends up to be abstruse for an outsider. For a comprehensive historical introduction to particle physics, underlining the main highlights, the reader is addressed to [Marshak(1995)].

Anyway, it is better to stop at this point: the more I add considerations on the subject of particle physics, the higher is the risk of writing nonsense. In the end, the goal of this presentation is only to offer a general description, emphasizing the role of photons in nature, within the framework of a self-consistent theoretical context. My approach, although still to be extended and validated, has the advantage of being coherent with classical results and a vision of reality in syntony with common understanding. The reader is obviously free to take into account the suggestions that better fit his own viewpoint and discard, after due consideration, that which seems meaningless.

2.6   Mass and gravitation

Let me devote this section to some observations concerning gravitational properties of particles. So far, we have used the term gravitation in relation to the presence of a modification of the space-time, in accordance with the theory of general relativity. Such a deformation is quite complicated, since in the case of bare particles it involves the support of a time-dependent electromagnetic wave, surmounted by a component able to confer a real mass to the particle (this is true for the electron at least, for the proton the construction can be far more complicated). Relatively to the electron, explicit computations show that the density of charge ρ and the pressure p, deduced by the solution of the modeling equations, can be used to recover plausible estimates of the effective mass at rest (see the definition of ρ_m given by (98) in appendix G and the successive computations in appendix H). It is known however that masses of elementary particles are considered to be too small to be responsible of the stability process of an atom or a molecule. I think that this viewpoint needs to be corrected. In truth, as we better check in chapter three, the stability of these systems is mainly due to their geometric displacement and to the pressure that the various components exert on each other. Such a mechanism must be somehow related to gravitation (in the broad sense of the term), and the fact that the measured mass of a particle is to some extent negligible does not belittle the importance of the geometrical setting.

Again I am noting that pressure p, and in particular its gradient, plays a crucial role in gravitational interactions, at the point that one is led to think that p can assume the meaning of gravitational potential (up to multiplicative dimensional constants), in a non relativistic sense. In a given gravitational environment, the term ∇p in (87) is proportional to forces that act only on regions where $\rho \neq 0$, as for instance inside charged particles. If this were true, it would be quite an elegant way to realize the so called unification of forces, one of the goals of past and present physics. In reality, physicists look for particle *carriers*, such as the *graviton* or the *Higgs boson*, but, as the reader may have already understood, my mind is set on a completely different approach. In principle, the unification of electromagnetism and gravitation is readily available as soon as one solves Einstein's equation[89](see also footnote 55). I have already shown how this can be done by putting the electromagnetic stress tensor on the right-hand side, and what interesting consequences can be drawn. The fact that, in a flat space-time, the "gravitational potential" directly appears in the constitutive model equations is far more important, since one can avoid solving Einstein's equation in order to recover gravito-geometrical information. Similar considerations were made in the case of free-waves (see section 1.6),

---

[89] From [Anderson(2004)]: "Wheeler's early hope was that vacuum geometrodynamics might turn out to be a Theory of Everything. [...] The properties of a distant isolated mass were argued to be recoverable from geons i.e. gravity waves almost completely trapped in some region (mass without mass). Aspects of charge could be recovered from the mere topology of empty space (charge without charge)".

by noting that electromagnetic fields already contain information about the organization of the geometrical part, even if in this case we have $p = 0$.

I am assuming that there is no distinction among gravitational and inertial mass. This principle of equivalence, as well as the possibility to transform electromagnetic energy into mass and vice versa, should be implicitly assumed at the same moment as writing my right-hand side tensor in Einstein's equation, thus linking the mass tensor with the metric space. The problem of identifying inertial and gravitational mass, well treated in [Fock(1959)], section 6.1, has however restricted validity, being limited to uniform fields and slow motions. Nevertheless, we learned that general relativity has a meaning that goes beyond the mere interpretation in terms of mechanical phenomena.

The novelty of my approach consists in mixing up in the energy tensor both the electromagnetic and mass contributions with different signs (see (94) in appendix G). These turn out to be well-balanced in the case of elementary particles, bringing to a neat characterization of the two competing components. The isolation of the bare electron from the rest is the first step towards a quantum view of matter. Without the help of differential geometry this path would not be viable[90]. For other types of unstable gravito-electromagnetic interactions, the percentage of the different ingredients is not defined a priori, but only obeys the global laws of energy and momentum conservation[91]. Anyway, before attempting any analysis concerning the concept of inertia, one should first know how to actually accelerate a particle. Let me postpone this issue until sections 3.1 and 3.2.

As I specified in section 2.2, the muon is an example of an isolated unstable agglomerate of rotating photons, perturbing the space-time geometry at an appreciable level, during its averaged life-time of the order of 10^{-6} seconds. In the context of my theory, the anomalous extra mass

---

[90]In [Fock(1959)], p.101, it is written: "It is natural to inquire into the deeper reason for the fact that in normal conditions the predominant part of the energy is bound so durably as to be in a completely passive state. Why does even a negligible part of it never leave this state and destroy the separate balance of the active part? The Theory of Relativity by itself is unable to answer this question. One should look for the answer in the domain of quanta laws, which have as one of their main features the existence of stable states with discrete energy levels". Contrary to what has been stated above, my belief is that, exactly because of the prerogative of general relativity to be able to suitably adapt the geometrical environment, the confinement of energy is effectively possible.

[91]From [Fock(1959)], p.101: "The particularly durable binding of the predominant part of all energy, or mass, is the reason why one can speak of the laws of conservation of mass and energy as two separate laws, although in Relativity the two laws coalesce into one".

of the muon is due to the presence of a further separation membranes between the bare electron boundary and the outside world. The membranes envelop a good amount of pressure intensity, justifying the presence of mass in addition to that of the contained electron. This precarious covering can be easily popped, setting free, through a discontinuous process involving a change of topology, the exceeding mass. In section 1.5, I mentioned that jumping from a free falling elevator to a standing floor is a discontinuous effect that produces a different perception on the measurements of weights. As a matter of fact, the change from the flat geometric environment of the elevator to that of the observer at rest is realized by passing through a bidimensional boundary, that is a separation membrane. The analogy between the elevator example and the motivation of the loss of mass of a decaying muon is very mild. However, it should enforce the idea that mass-energy transformations occur during sudden changes of topology, caused by the rupture of interface boundaries. Information about mass can then be stored on confinement membranes and quantitatively encoded by measuring surface tensions (see section 2.2). Pressure discontinuities are actually concentrated on certain surfaces. A similar approach has been considered, in a totally different abstract context, by other authors[92]. In proton decay, due to collision with anti-protons (see figure 2.10), things are more complex but substantially analogous. We must not forget that conservation of energy and momentum are always assumed, and after any transition process, mass is transformed into electromagnetic energy and vice versa, and this is done in the search of the most stable equilibrium and in the direction of increasing entropy.

Once again, let me recall the main ideas. The scalar p is constantly zero when the corresponding photons behave as free-waves, i.e., they travel with constant uniform speed equal in intensity to c. As soon as a photon is subject to external factors (that can also be self-interactions, as in the case of the system of circling waves in a toroid), then the gradient of p starts to be different from zero, so generating non-trivial pressures. This agrees with the mechanics of bodies, where a change of trajectory is related to the action of forces. Kinetic energy is thus converted into potential energy and vice versa. Thanks to Einstein's equation that ensures conservation of energy and momenta, the total balance is always preserved. Note also

---

[92]From [Verlinde(2011)]: "Thus we are going to assume that information is stored on surfaces, or screens. Screens separate points, and in this way are the natural place to store information about particles that move from one side to the other".

that the fact that light may exert pressure on objects[93] (the case of *optical tweezers*, for instance) is a known phenomenon, that is not easily explained by quantum theories but is naturally embedded in the framework of my theory. Therefore, for given initial data, it is enough to solve the system of model equation in a prescribed metric space (the flat one for instance), in order to know both electromagnetic and gravitational type behaviors. As the accelerations end, pressure and its gradient go back to zero and the various components proceed with uniform constant velocity, i.e., they behave as free electromagnetic waves following the rules of standard geometrical optics. Pressure may however be trapped in "bubbles", that can last forever (as in the case of stable particles) or for short periods of time. To realize this, there must be an instant when the particle container, initially opened, is suddenly sealed: charge (unitary) and mass are thus confined.

The crucial question is now to see if p may be quantitatively regarded as a gravitational potential. According to my construction, by examining the constant χ in front of the energy tensor T to be placed at the right-hand side of Einstein's equation (see appendices E and G), we find an inconvenience: the associated "gravitational constant" $\tilde{G}$ is predicted to be about 10^{41} times larger than the standard one, usually denoted by $G \approx 6.67 \times 10^{-11}\,\mathrm{m}^3/\mathrm{s}^2\mathrm{Kg}$. As a matter of fact, the most popular choice for the constant multiplying the mass tensor is $\kappa = 8\pi G/c^4$, while in my case we find $\chi\mu^2/c^2\epsilon_0$, where $\chi \approx .15$. Considering that $\mu \approx 2.8 \times 10^{11}$, the ratio $\tilde{G}/G$ turns out to be approximately 5×10^{41}. It is then necessary to comment on this result and possibly come out with some explanations.

The classical Newton law of gravitation seems in any case useless at atomic level[94]. Let me first point out that this is partially true also at cosmological level, where it is common to distinguish between the gravitational and the geometrical radii of a given mass. For instance, in disagreement with celestial observations, standard exact solutions of Einstein's equation require the predicted gravitational radius to be much smaller than the effec-

---

[93] From the review paper [Cattani et al.(2013)]: "We have thrown a look at one of the most exciting developments in physics. The possibility of manipulating matter with light, which is not an ideal theory anymore thanks to sources of both coherent electromagnetic radiation and coherent matter waves".

[94] From [Hawking and Israel(1987)], p.15 (S. Weinberg): "We have a theory of gravity, Einstein's theory of general relativity, which reduces to Newton's theory at large distances and small velocities. This theory of gravity works very well on the scale of the solar system or the galaxy or here on the scale of everyday life on the surface of the earth, but it is a theory which when pushed to very short distances and high energies begins to give mathematical nonsense".

tively measured geometrical radius, and the discrepancy gets more evident as masses are larger. This is true for instance for the *Schwarzschild metric*, recoverable from (63) (appendix E) by plugging $Q = 0$. On the other hand, Newton's law of gravitation does not give any prescription about the effective size of the masses involved (one must know density, but density depends on the context).

I recall that the gravitational radius is given by $2Gm/c^2$, where m is the mass of the object under consideration. For example, according to the Schwarzschild solution, if the Earth was a *black hole* of equivalent mass, its diameter would be less than a centimeter. In this guise, despite such an assumption on the Earth's size not being realistic, the asymptotic behavior of the gravitational field at large distances from the Earth would be in agreement with Newton's law. By extrapolating, if we assume the proton to be a black hole having a diameter of 10^{-15} meters and a mass of the order of 10^{-27} Kg, general relativity would suggest a gravitational constant 10^{40} bigger than G, in agreement to what has been found in the paragraph above. The reader must admit this observation is rather astounding. Note that my idea of elementary particle matches that of a black hole, interpreted as a massive object from where light cannot escape, since this is what really happens when photons are constrained in that tiny portion of space. It is however a far more complex structure than those studied in cosmology, where for the sake of simplicity bodies are often reduced to spinning spheres. Once again, let me also point out that with a switch of the sign of the right-hand side of Einstein's equation (maintaining the signature of the space) it is possible to relax the constraint on the horizon of the *Kerr–Newman* metric (see [Funaro(2009b)] and appendix E), granting the theoretical existence of relativistic charged singularities with very small masses.

Astronomers may consider this argument rough and superficial. It is probably true, but one cannot deny that there are open questions regarding the identification of real masses and certain well-known solutions of Einstein's equation. Thus, standard relativistic arguments are also affected by a good dose of naive passages. In truth, although very celebrated and ubiquitous, the Schwarzschild solution and its upgraded variants, can be considered reliable only at large distances and small velocities, while their behavior at different regimes is going to be totally disjoint to questions properly related to gravity[95,96]. In addition, such a solution represents the gravitational field produced by a singular point-wise stationary mass, that is more or less analogous to simulating a point-wise stationary electric charge in the framework of Coulomb's law, having the *Laplace operator* equal to zero at all points with the exception of the source. Charged particles are far from being so simple, and I have already pointed out in many circumstances the limits of electrostatics. Therefore, for my purposes, the Schwarzschild solution offers an oversimplified description. The forcing right-hand side in the Schwarzschild case is zero everywhere except for a single point. It is quite a borderline pathological situation. I believe therefore that major attention should be devoted to smooth solutions.

Classical and relativistic approaches are closely linked but quantitatively separated by a series of approximations and neglected terms. The models available, although similar in some respects, offer in the end an incomplete picture. The situation gets even more unpredictable when masses below a certain magnitude are taken into consideration. The gravitational constant G remains in fact one of the most difficult quantities to estimate and, in the case of relatively small bodies, measurements can be heavily influenced by several other factors. The aim of the above specifications is to point out how insidious is the matter at hand.

Another crucial aspect of Newton's law is that it deals with masses, independently of their nature and density. This could be a limit for small objects. Indeed, when we examine a molecule, its mass is very small on the whole; nevertheless, it is concentrated on the nuclei and, in smaller part, on the electrons. In between there is imponderable photon energy (to be studied in chapter three). The spacing between nuclei is of the order of 10^5 times their size, that, in terms of volumes, amounts to 10^{15}. Thus, the mass density of a nucleus is of the order of 10^{15} times that of "normal matter". If there was the possibility to adapt Newton's law in order to include an indication of the densities of mass of the objects involved, an updating of the gravitational constant would be necessary. The new constant $\tilde{G}$ is going to be much larger than G, returning to the standard G for large

---

[95]Radial geodesics of the Schwarzschild metric are computed in [Atwater(1994)], section 4.3. In particular formula (4.31) tells us that, for non negligible velocities $(dr/dt > ac)$, the gravitational field of a positive mass is repulsive.

[96]From [Sachs(2004)], p.64: The theory proves that in the Newtonian limit of general relativity, the force of gravity can only be attractive (an empirical fact). But, when we are not in this approximation, the force of gravity can be either attractive or repulsive, thereby leading to the dynamics of the oscillating universe cosmology.

aggregations of particles. In addition, we can explicitly check that if a new $\tilde{G}$ were 10^{40} times larger than G, there should not be much difference in magnitude between the repulsive electric force of two protons and their gravitational attraction. Having pointed this out, one may start realizing that such infinitesimal gravitational masses, as that of a proton, are not so small after all, if considered in relation to effective sizes and densities. In fact, the standard constant G is based on an averaged mass of a body, consisting of a multitude of spikes. As one enters the lattice structure, the corrected constant $\tilde{G}$ must be used. The solutions prospected above are just palliative strategies, since, as it will be stressed in the sequel of this exposition, the real gravitational effects develop at the interstices between particle. When approaching a piece of matter, force fields start mixing up without necessarily passing through violent jumps, so that one cannot quantitatively distinguish among the various contributors. The mixture can produce fields of new flavors, such as *Van der Waals* forces[97]. To tell the truth, there are no such different effects in nature, but everything descends from the apotheosis of electromagnetism.

The real fact is that gravitation cannot be coded in simple rules, at least not so simple as the Newton's law of attraction. Let me specify that I totally trust the theory of general relativity, and I do not believe it needs further adjustment, although alternative proposals or improvements have been copiously proposed. I dislike however the direct connection between general relativity and the empirical laws of astronomical gravitation, which are just imprecise descriptions of a complex system. General relativity was validated when the so-called precession of the planet Mercury, theoretically predicted, was actually confirmed. Such a property could not be deduced through previously existing laws. Nevertheless, this is another example of astronomically related achievement, that depicts Einstein's theory as a completion and rationalization of results in that specific area. It is an underemployed use of a tool that is very rich in potentialities, able to describe phenomena that go far beyond gravitational effects. It has to be said that the theoretical and numerical study of Einstein's equation presents nontrivial difficulties. However, nowadays the research seems to have plateaued on

---

[97]The molecular interaction zoo is described for instance in [Sharpe(1981)], p.85: "... it is customary to divide the types of bonding into ionic, covalent, metallic and van der Waals's (hydrogen bonding is sometimes added as a special category). The weak residual attraction between molecules which are not covalently bonded, for example, are often grouped together under the heading of van der Waals's forces, though it is now recognized that three types of interaction are involved: dipole-dipole, dipole-induced dipole, and dispersion (or London) forces".

a certain number of cosmological problems, and, as frequently happens, extending simple solutions outside the domain of their validity is an unstable process, sometimes leading to conclusions that explain both everything and its contrary. I think I have already provided here plenty of examples where general relativity could play a significant role, without being relegated to gravitational facts. In other words, in the micro-world, gravitation is of fundamental importance, but, in this case, it it not merely the completion of a set of rules based on the attraction of bodies. It represents instead a unifying theory, able to deal with forces at various scales, manifesting themselves in several ways, both in the attractive and the repulsive regimes (see footnote 95). Current research at the nanoscale level offers new vistas.

In many cosmological problems, the universe is equipped with some background metric, which is usually not the flat one, where masses are given and evolve, further modifying the geometrical setting. One of the questions is to understand what is the real form of the universe, with special attention to experiments showing its mode of expansion. Although a general behavior could possibly emerge, my guess is that the geometrical description of our universe is the sum of an immense number of dynamical events, each one carrying its own contribution. Critically examining the existing literature, other authors expressed their positive opinion about the necessity to furnish the universe with a non homogeneous dynamical background, roughly defined as aether for lack of a more concise formalization[98,99]. The background is a continuum that links various structures[100], but the two

---

[98] From [See(1920)]: "... the existence of forces implies stresses in the aether: the stresses imply waves: the waves imply heterogeneous density in the medium, which must vary with the radius from any mass according to the law $\sigma = \nu r$. There is no other view of the aether which can be held. Homogeneity of density would imply no stresses: no stresses would imply no forces; no forces would imply an inert universe; which is contrary to observation and thus wholly inadmissible. To fix upon a more familiar everyday image of this world structure, we may imagine a box filled with large oranges, and the finest dust, like that of lime, or smoke from a cigar, penetrating the relatively vast spaces between the oranges, which however should not be in contact, but in rapid motion. If now the cigar smoke, or the particles of lime dust, be imagined to have stupendous velocity, flying hither and thither with inconceivable speed, and thus moving with the utmost freedom in the open spaces between the oranges, as well as outside of them, we shall have a very good image of the behavior of the aether in respect to matter".

[99] From [Minini(1985)], p.172: "L'ultima conclusione è perciò molto semplice: potranno esservi in questo studio idee inesatte o del tutto errate, molti errori, ma ciò che è certo e che deriva logicamente dall'esame delle idee scientifiche di questo secolo è che esiste un etere in movimento all'origine di tutte le forze dell'universo".

[100] From [Sachs(2004)], p.149: "The single theory that we examine, then, with physical implications in all domains of nature, from the microscopic domain of 'particle physics'

aspects are not independent; they actually have the same electromagnetic origin: matter can be destroyed becoming part of the background, energy of the background can give birth to matter. In addition, there are unstable zones where this process of transformation is not concluded; energy there is neither fish nor fowl, presenting mass density that does not evolve in concrete mass. It is impossible to understand nature by viewing it as a crude mechanical interaction of lonely subjects, because things are unavoidably interconnected and, at certain scales, the holistic vision is the only way to interpret their secrets.

The construction sequence starts at the atomic level, but at certain stages of development the geometry can be mediated by larger clusters[101]. Therefore, it is not possible to refine the knowledge at the upper level, if one does not take into account the myriad of intermediate passages. Moreover, the phenomenon of gravity is continuously changing with time, involving a full range of frequencies from that of gamma rays to that of the slow motion of galaxies. For these reasons, I think that the term gravitation, as it is commonly adopted nowadays, only refers to an averaged description of large-scale behaviors, but misses the opportunity for studying complex molecular phenomena, very rich in dynamical substructures, where geometry is the codification key. To better appreciate this scenario and the links between the different passages, additional comments are provided in section 3.8.

Finally, it has not to be forgotten that, according to my opinion, the source of gravitational signals is electromagnetism, the primary building block. Untying the secrets of electrodynamics is the doorway to understanding gravity. Concurrently, space-time deformations drive electromagnetic signals along specific paths. Kinematic energy turns out to be an inseparable mixture of electric and magnetic contributions. The conversion into potential energy is the start of tangible gravitational effects, that

---

to the domain of cosmology - that of the universe as a whole - is the theory of general relativity. Its ontological stand is different from that of the quantum or classical (Newtonian) theories of matter. In this view, then, the universe is not a collection of separable 'things', interacting with each other or not. Instead, the universe that the scientist investigates is a single holistic continuum, with an infinite variety of (nonseparable, correlated) modes. It is the modes of the universe that appear to us as its constituent things - such as the galaxies, individual stars and planets, people, trees, electrons and protons".

[101] From [Oppenheimer(1989)], p.40: "These acausal atoms compose the familiar world of large bodies, orbits, and Newton's laws. The laws that describe atomic behavior, the stationary states and transitions, reduce by correspondence, when applied to large systems, to Newton's laws".

may culminate in the formation of stable particles. Here, the symbiosis is high. Evidence of this process is the mutual, instantaneous and synchronous creation of both charge and mass, as far as simple stable particles are concerned. By the way, the study of astronomical interactions often neglects electromagnetic waves, since they seem not to have a decisive role. I stated that gravitational waves can be associated with electromagnetic ones via Einstein's equation. Gravitation turns out to be then a pure and entire consequence of electrodynamics interactions. Besides, it has also been observed that the amplitude of gravitational waves is inversely proportional to their frequency (see section 1.6). As a result of this important remark, very low-frequency electromagnetic signals may carry significant contribution at gravitational level. Nevertheless, these phenomena evolve at the speed of light, which means that their detection is almost impossible on the medium-small scale. Gravitational effects at low-frequency are going to be a by-product of electromagnetic waves of a very small intensity and variation that nobody really cares or has the means to measure.

On the other hand, a good deal of astronomical research is closely connected to magnetic phenomena. To quote an example, one may think about the terrestrial magnetic field and the the famous *van Allen belts*, that, not incidentally, display a toroid shape. *Relativistic Magneto Hydrodynamics* is a branch of physics that, in the framework of general relativity, merges the constitutive equations for a conductive fluid with contributions due to magnetic sources, in a fashion very similar to that described by my model.

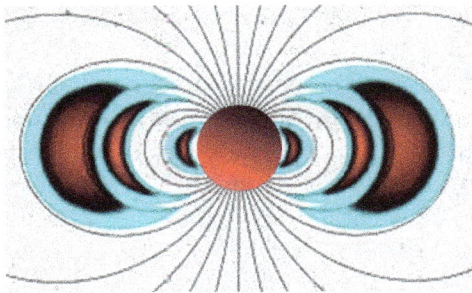

The context of applications, as well as the consequences, are however very different from the ones here considered. In chapter three, I will claim that our universe is filled with a colossal amount of fluctuating and organized electromagnetic energy, also contributing to generate and transfer gravitational information. This implicitly means for instance that the Earth is at the mercy of intense, although hardly detectable, gravitational oscillations brought by extremely low frequency electric and magnetic perturbations. If it is true that electromagnetism is the primary source of gravitational forces and what I am presenting here is the decryption key, one should not be too far from setting up experiments aimed at the production of mechanical effects from electromagnetic devices (with this I mean something fancier than a classical electric engine). To this end, one may take into account that, as light rays are plied by gravity, the other way round, a sudden variation of trajectory imparted to photons activates the mass tensor at the right-hand side of Einstein's equation, with the consequent production of pressure; and when the presence of pressure is registered it means that one has entered the world of mechanics. Some theoretical considerations, concerning the possible generation of gravitational waves from accelerated electromagnetic ones, have been made for instance in [Ardavan(1984)]. Scattered, non officially validated, experiences report the possibility of obtaining "gravitational shielding" from fast rotating apparatuses presenting peculiar electrical properties. An interesting subject is the possibility to launch pressure waves towards faraway regions. Some experiments (see [Podkletnov and Modanese(2003)], [Poher and Poher(2011)], [Modanese(2013)]), based on the high-voltage discharge of a superconductive cathode, actually generate radiation pulses of non electromagnetic type that seem to act as gravity repulsion beams, as testified by acceleration sensors placed at some distance from the source. These heavy "space-time" deformations (can they be named gravitational waves?) have been registered to travel at superluminar velocities. Are they sustained by longitudinal electromagnetic pulses? Or they can autonomously proceed in the vacuum? The adoption of a superconductive material for the electrode gives to the discharge spark an uncommon distributed shape, with the result of reaching the anode in a non-rectilinear way. We may have all the ingredients for guessing some explanation in accordance with my viewpoint.

Chapter 3

The Constituents of Matter

Millenni di sonno mi hanno cullato
ed ora ritorno. Qualcosa è cambiato
non scorgo segnale che annunci la vita
eppure l'avverto ci son vibrazioni
Franco Battiato, composer

3.1 The Hydrogen atom

After having devised a preliminary sketch of electrons, protons and nuclei, the next situation to be handled is the formation of the simplest atomic structure: the Hydrogen atom. Recall that, up to this moment, we have mostly dealt with bare particles, analyzing their internal structure and what happens in their vicinity. Thus, we need something in between to allow more incisive communication between the parts, and since I only have electromagnetism in my menu, I will play again with photons. From now on, my exposition will be even more qualitative. Although direct numerical simulations can be performed on the mathematical differential model, such computations look rather intensive and need careful preparation. I will base my arguments on intuition and experience. I basically treat electromagnetism as a fluid and take inspiration from mechanics to describe its behavior.

In chapter two, I claimed that the energy surrounding a bare particle has chances to be dragged into a toroid shaped envelope (see figure 2.2). It should be clear that, whenever the conditions are favorable, the process can be repeated. As long as there is energy floating around, this is driven to form a matrioshka-like sequence of encapsulated photon shells. It has to be remarked that at the interface boundaries there is a discontinuity of the velocity field. It is required however that the arrows, even if they display different intensities, are directed in the same way (opposite verse instead for the case of matter and anti-matter junctions). These containers hereafter will be called *photon shells*; the term is applied to general closed configurations of various shape and topological organization, that can also be totally disjoint from bare particles.

Before continuing, let me recall what we got in the previous chapter. The bare electron is a classical ring with photons circulating in it. In the first picture of figure 2.7 such a ring is surrounded by a first shell. Successive shells can be added in similar fashion. The shells do not carry any discernible charge (the fields oscillate with a time-average per period equal to zero), however stationary electric terms with $\rho = \mathrm{div}\mathbf{E} = 0$ and $\mathrm{curl}\mathbf{E} = 0$ may be present. One can match things in such a way that the whole electrostatic solution (i.e., the stationary part of the electric field), defined both on the bare particle and the subsequent shells, is globally continuous (see figure 2.2). The conditions on $\mathbf{E}$ imply that the electric field decays as the inverse of the square of the distance from the bare particle. Recall again that inside the particle one has $\mathrm{div}\mathbf{E} \neq 0$; this prevents field singularities. I also examined the possibility that the first photon shell encircling the bare particle is such that $\mathrm{div}\mathbf{E} \neq 0$. This is the first step to create massive particles of increasing complexity (see the second picture of figure 2.7 and, in section 2.3, the reasons given in support of this construction). This possibility is not taken into consideration in the present chapter.

Protons are much more complicated. Basically, I assumed that they are constituted by an electron-type ring immersed in a secondary photon shell containing other spinning photons (see figure 2.9). In this system we globally have $\rho \neq 0$, so that the structure acquires both charge and mass. Differently from the case of the electron, the case of the proton is a bit peculiar. The external shells can be numerous, and their influence can be felt at large distances. Energies associated with shells decay quite fast and this is also true regarding their frequencies of rotation. On the contrary, number and magnitude of the shells in the neighborhood of an electron depend strongly on the environment and will be analyzed in more detail later. For simplicity, the bare proton has been represented as a (almost spherical) Hill's vortex, i.e., a vortex with the inner hole degenerated into a segment. As far as the successive shells are concerned, the reader can refer to figures 3.1 and 4.5. The study of 3D electromagnetic waves, solving all of Maxwell's equations and trapped in perfectly spherical Hill's type vortices, has been treated in [Chinosi, Della Croce and Funaro(2010)], case D, both analytically and numerically. The dynamics is analogous to that sketched in figure 2.9. Let me specify that these cases are hypothetical. Other shapes could in principle be devised, though the expected dynamical behavior should not be dissimilar. One has also take into account that the framework is not rigid, since it may change under the action of external mechanical forces; those that later will contribute to the vibrational properties of molecules.

The reason why these shells develop could be due to the oscillating properties of the electric field at the separation surfaces. There, the magnetic field is zero and **E** vibrates tangentially. Such oscillations are not globally synchronized on the whole surface, but they activate sequentially according to the orientation of the velocity field **V**. Like infinitesimal dipoles, these sources transmit their message by following a complex scheme, difficult to explicate without the help of numerical simulations.

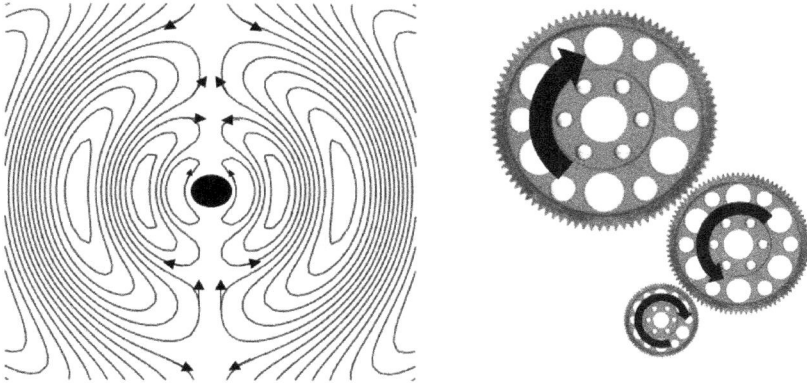

Fig. 3.1 *Differently from the electron, a proton generates a sequence of encapsulated photon shells, whose radius grows geometrically. In each shell there are spinning photons, synchronized with a phase difference and organized to form toroid structures. The direction of rotation changes by passing from one shell to the successive one. The situation recalls that of a series of gears of increasing size. By turning the small gear the rotation is transmitted to the larger ones. The photons follow a similar motion, although the contact is not point-wise as in the case of the gears.*

The rules of electrostatics will continue to hold, and point-wise charges are substituted by charged dielectrics with annular shape. There is a privileged way to prolong the solution compatibly with the dynamic part, and this is amenable to a Coulomb-type potential centered in the particle. As a consequence, fields present at the outer surface of each shell turn out to be uniquely determined by those given on the inner boundary (going backwards, everything originates from the bare particle). As it will be shown later, this may be a source of conflict when shells related to two different particles meet. Let me also say that, since $\rho = 0$ outside the bare particle, the vector $\mathbf{V}$ is not necessary and the pressure p can be taken to be equal to zero (see equation (87) in appendix G). As a consequence, the shells do not display any mass. Nevertheless, we may define $\mathbf{V}$ by recalling that it has been introduced with the aim of signaling the direction of the electromagnetic energy flow. The velocity field $\mathbf{V}$ is not associated to the actual movement of physical entities, but, like in a *Mexican wave*, to the way the information is transmitted. In this fashion, a direction of rotation is associated to each shell, regardless of the effective local development of the dynamical fields. Based on figure 3.1 there is a common axis of symmetry, so that the notion of spin orientation of a particle can be transferred to the external shells.

At least in the proton case, by the mechanism of developing shells (if neighboring energy is available), the particle can transmit information about its charge far away, and this is probably done at velocities comparable to that of light. Let me point out that, with the classical approach of considering the particle as a point-wise singularity subject to Coulomb's law, the transfer of information is immediate, i.e., isolated charges at enormous distance feel their reciprocal existence instantaneously. Although a large community of physicists support the idea that some interactions may occur with almost zero propagation time, perhaps because of connections through other space dimensions, I would prefer to remain in a "classical" 3D universe. Thus, an entire proton consists of the bare core and a possibly unbounded sequence of photon shells. Hence, the velocity of propagation of the electric information depends on how rapidly the shells develop, and, in any case, does not pass the speed of light. The frequency associated to each shell reduces as one moves from the core, proportionally to the inverse of the distance (see a few paragraphs ahead). The last observation is very crucial in my theory, since my interest in atomic physics started with the aim of generating such an effect, that is a decay of the frequency with distance (see section 3.8). As in the case of real non-viscous fluids, conservation of energy and momenta dictate the laws ruling the various shells. An analogy can be established with a set of communicating gears of different sizes (see figure 3.1); a tricky appliance that finds roots in Maxwell's mechanical description of electromagnetism (see footnotes 42 and 43).

It is my opinion that the role of principal actor, in the transfer process from a proton to the external world, is played by the magnetic field, for the reasons that I will try to explain in the coming paragraphs. As stated at the end of section 2.2, a stationary magnetic component is present strictly inside the particle, which is not prolonged to the outside. The dynamical component however remains in conjunction with the electric one, as the classical electromagnetism prescribes. We know that in the Maxwellian context, a wave equation for $\mathbf{B}$ holds (see (6) in appendix A). I will also make the crucial assumption that the frequency of rotation of the photons in each shell decays as the inverse of the distance from the proton. For the arguments put forth in section 3.8, such a change of frequency is not made in continuous fashion, but follows a quantized behavior, which is the one actually responsible for the creation of different shell layers. Combining together the two ingredients (information is transferred through oscillations and the frequency decreases), I propose in appendix I a differential model in vector form, the construction of which follows from heuristic arguments. I called $\hat{\mathbf{B}}$ the solution of such a stationary eigenvalue problem, although now there is no quantitative connection with the magnetic field $\mathbf{B}$. The idea is that the border between a shell and a contiguous one corresponds to the level surfaces where $\hat{\mathbf{B}} = 0$. The first of these surfaces is the spherical one associated with the proton skin (corresponding to zero boundary conditions). Note that $\hat{\mathbf{B}}$ is determined up to a multiplicative constant; that, by the way, does not affect the domain of the zeros.

Computations that take into account the size of the proton (about $10^{-15}m$) and the *fine structure constant* $\alpha \approx 1/137$, show the existence of a series of encapsulated spherical shells. Their size grows geometrically and their associated frequency decays accordingly. This property of *self-similarity* (that a geometrical sequence can be transformed into itself by a suitable multiplicative factor) is a first contribution to the *fractal* structure of matter, that will emerge as I go ahead with the discussion. Two shells are the most peculiar. The first one is relatively small (just about 400 times the diameter of the proton). I guess that, if it were real, its existence should have been noticed. In truth, in the history of low-energy elastic scattering, there is reference to a mysterious particle called *pomeron* (see, e.g.: [Low(1975)], [Nussinov(1975)]), responsible for unexpected behaviors in experiments aimed to determine the proton cross-section. It would be interesting to better investigate this analogy. The second characteristic shell has a diameter of the order of one Ångström (1 Å= 10^{-10} meters), which is a typical size for molecules in general. On such a shell surface we have $\hat{\mathbf{B}} = 0$, and also the magnetic field should almost vanish (remember that $\mathbf{B}$ and $\hat{\mathbf{B}}$ are different things, the first one is a measurable field, the second one is a surrogate introduced to simplify the calculations). This is the "quite zone" where I expect to find electrons in equilibrium. I will soon return on this crucial point.

Let me observe that together with the charge and the magnetic information, also the geometrical properties (including spin orientation, as said above) are transferred through the shells. This means that also the gravitational message is spreading out. In this context, "gravitation" assumes a very general meaning that goes far beyond that commonly ascribed for instance to astronomy. Let me recall some concepts already expressed in the previous chapter. Nuclei, atoms and molecules are essentially kept together by gravity, despite the popular argument claiming that, since particle masses are negligible, there is no reason to involve gravitational issues at the atomic level. These statements certainly sound very eccentric, but my purpose here is to slowly bring the reader to my side and let him appreciate the advantages of my approach. For this, we have to subvert the order of things and interpret well-known experimental results in a new spirit. This implies an entire review of some consolidated assumptions, without denying the evidence of the available data.

I mentioned in the introduction that the easiest explanation of a phenomenon is not necessarily extendible to all circumstances (see also footnote 174). As remarked in section 2.6, applying the rules of standard gravitation obtained for large mass bodies is not straightforward in the microcosm. For this reason, it is unfair to use the usual inverse squared law, introduced to study the attraction of two planets, in the limiting case of two particles. As a matter of fact, it wrongly leads us to believe that gravitation is unnecessary at those scales. Stable elementary particles are provided with the same features and flavors of all the objects we find in our universe. Neglecting gravitational effects sounds like a non-optimized use of resources. Thus, I will be very radical in this respect. Following the pattern of previous chapters: electromagnetism relies on photons, photons are geometrical emanations, geometry is the embryo of gravitation, gravitation governs electromagnetism. Geometry may keep photons together, showing that particles are glued by "gravity". Some electromagnetic energy may be missing but only because it has been converted into some gravity potential energy (by meaning that it has potentiality to do work), ready to be transformed back into photons. Therefore gravity, intended as a dynamical geometrical process is also the cement of atoms, and I am going to discuss how the miracle actually happens by starting with Hydrogen.

Physicists and chemists seem to be satisfied with the current model of atoms, and the technological world has proliferated based on it. The search for a justification that goes beyond quantum mechanics dogma has been abandoned by most. I always thought that, although practically effective, the explanation of the structure of matter given by official sources is totally unsatisfactory and leaves a sensation of incompleteness[102]. Sometimes definitions grasp at straws[103].

Let me naively review the main steps starting with the *Bohr model*. In the Bohr Hydrogen, the electron physically rotates around the proton. This vision soon generates questions regarding the choice of a preferred observer frame, but let us forget this for a while. The representation is so rooted in popular iconography that atoms are depicted as a central granular bunch encircled by trajectories of spinning electrons, and when asking a friend to describe Hydrogen he will involuntarily start moving a finger in rotatory fashion. The electrical attraction of the two bodies maintains the orbit. Nevertheless, not all the trajectories are possible.

Due to the *de Broglie* interpretation, the electron is also a wave, carrying a frequency proportional to its momentum. It turns out that there is an in-

---

[102] From [Hawking and Israel(1987)], p.26 (R. Penrose): "We are here confronted with the odd picture of reality that quantum mechanics presents us with. [...] Some people feel uneasy with such a clear-cut geometrical pictures of quantum states. They say that we should not try to form a picture of reality when applying the rules of quantum mechanics: just follow the rules of quantum mechanical formalism, do not try to form pictures and do not ask questions about reality! This seems to me to be wholly unreasonable".

[103] From [Bohm and Hiley(1982)], p.236: "In the hydrogen atom the electron does not appear as a wave or as a particle, but in a form different from both, namely, as angular-momentum and energy eigenstate, in which it has neither a definite position (particle) nor a definite momentum (wave), but does have a definite angular momentum (rotator). The classical picture that come closest to this is that of a standing spherical wave".

finite set of quantized "states", such that the electron's frequency, induced by the velocity of rotation, agrees well with the corresponding orbit frequency. These can be exactly evaluated, by knowing a few basic constants. The kinetic energies resulting from this computation adequately match observations, and this, at the time of the discovery of such a primitive atom model, was the first significant validation. Though simple and appealing, this idea is not trivially adaptable to more complex atoms[104] (not at all, I would say). In the case of a molecule it is very hard to imagine how a structure can be stable with all those staggered circumnavigating electrons. It has to be taken into account anyway that mechanical interpretations have been, and still are, sources of inspiration[105,106].

In the successive and "upgraded" explanations, critical problems have been disappointingly hidden. Through the work of W.K. Heisenberg, quantum mechanics claims that determining the exact position of an electron is an ill-posed question. The calculation is reduced to the determination of a distribution of probability, having higher intensity where there are more chances to find the electron. The proton is then supposed to be covered by a cloud of expectation values. Schrödinger's and Dirac's theories provide us with an effective means of calculating the "correct" distributions, passing through the resolution of suitable differential equations and the evaluation of the associated eigenfunctions. The corresponding sequence of eigenvalues gives us a glimpse of the energy distribution and agreement with reality is now achievable even for more complicated atoms and molecules (not too complicated anyway). Further extensions take care of the concept of spin and *Pauli's exclusion principle*. By the way, I do not want to proceed with this discussion any longer. I believe in a complete causal description of nature phenomena, and I consider the statistical approach to be a contingent result, temporarily set up to collect the main ideas, but still waiting for

---

[104] From [Miller(1984)], p.129: "... Alas, continued Born, the 'possibility of considering the atom as a planetary system has its limits. The agreement is only in the simplest case [the hydrogen atom].' The honeymoon of the Bohr theory was over".

[105] From [Bokulich(2008)], p.130: "Although it is typically believed that classical trajectories were banished from quantum systems with the downfall of the old quantum theory, recent work in semiclassical mechanics reveals that they still have a legitimate, though revised, role to play".

[106] From the Preface in [Dragoman and Dragoman(2004)]: "In contrast with the mainstream of accepted wisdom, we consider that quantum-classical analogies are a source of understanding and further development of quantum physics. Indeed, many quantum physical concepts have originated from classical notions, a striking example in this respect being the Schrödinger equation, which was formulated by starting from classical optical concepts".

radical reorganization[107,108]. Therefore I would prefer to stay within my set of deterministic equations, where the solutions are exactly telling me what happens time-wise and point-like. I am inclined to accept most of the hints and conclusions of quantum mechanics, but I also aimed to disclose the underlying machinery.

Let us go back to the Hydrogen atom. How do we check that a proposed model fits experiment? We know that it is possible to associate to the atom a multitude of different quantized energy states and that jumps between one state and another may occur. These changes are characterized by the emission or the absorption of photons, globally having the same amount of energy as the gap between the two Hydrogen states involved. Well, well, there are photons inside Hydrogen! Trapped photons. They indeed provide part of the atom's energy. When abundant, some of them can be ejected, leaving the atom in a lower energy state. Conversely, pumping some form of energy into the atom corresponds to adding photons.

And there are also frequencies involved. The *spectrum* of Hydrogen is a set of prescribed and very characterizing resonance frequencies[109], that can be easily pointed out by tickling the atom with some source (note that some frequencies are in general outside of the visible range). This is a sign that those trapped photons are "rotating". Thus, something is actually circulating inside Hydrogen, although it is not the electron. So, let the light photons do the job that cannot be undertaken by massive electrons. Indeed, in my eccentric model of the atom there is no movement of particles, that are

---

[107] From [Bohm(1957)]: "It is possible that looking to the future to a deeper level of physical reality we will be able to interpret the laws of probability and quantum physics as being the statistical results of the development of completely determined values of variables that are at present hidden from us. It may be that the powerful means we are beginning to use to break up the structure of the nucleus and to make new particles appear will give us one day a direct knowledge that we do not now have at this deeper level. To try to stop all attempts to pass beyond the present viewpoint of quantum physics could be very dangerous for the progress of science and would furthermore be contrary to the lesson we may learn from the history of science".

[108] From [Gross(2005)]: "Einstein, who understood better than most the implications of emerging interpretations of quantum mechanics, could never accept it as final theory of physics. He had no doubt that it worked, that it was a successful interim theory of physics, but he was convinced that it would be eventually replaced by a deeper, deterministic theory".

[109] From [Oppenheimer(1989)], p.41: "Whenever light acts on matter, or is produced by it, we find packets of defined energy and impulse, related to their frequency and wave number by the universal proportionality of the quantum of action. How were these quanta to be thought of? Were they guided by the waves? Were they the waves? Were the waves an illusion, after all?".

instead joined through a tumultuous circulation of trapped photons. This construction is very obvious and represents my modest Copernican revolution. Those in search of more fancy explanations will remain disappointed. On the other hand, with the following ingredients: nucleus, electrons, photons; this is the best recipe I can come out with. It is necessary anyway to justify in a convincing way all the passages.

First of all, optimal configurations for Hydrogen are supposedly obtained when the spin axes are lined up. Using terms borrowed from the dynamics of fluids, proton and electron attract each other as two little vacuum pumps, and stay together by virtue of some *Venturi*-like effect. As in the *Casimir effect* explaining the anomalous attraction of two uncharged metallic plates (I will return to this issue later with more details), there is more energy outside the system of the two particles than in between, and this generates the pressure that keeps them together. The issue is indeed more delicate and its explanation must be improved upon. In addition, we do not have to forget that there is a minimum distance from proton and electron and this turns out to be mainly due to the presence of the intermediate components. This avoids the mutual collapse of the two particles and guarantees the stability at the ground state.

Let us examine what happens with more attention when two particles interact. The remarks I am going to make are decisive for the comprehension of more general situations. Note first that two protons (or two electrons) will never be able to realize an equilibrium. Everybody will promptly say that this is evident from the fact that they display charges of the same sign, but, if we discard the action-at-a-distance argument, an alternative explanation must be given.

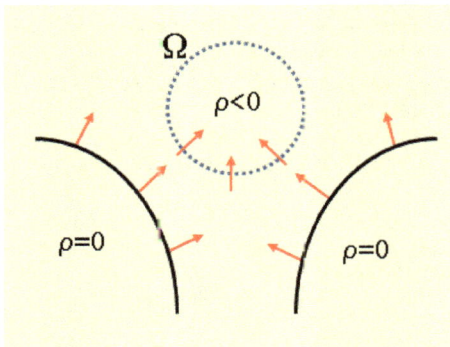

Fig. 3.2 *The electric fields present at the surfaces of two independent photon shells, relative to two distinct particles of the same charge, give raise to a region Ω where $\rho < 0$. Pressure increases in Ω and the shells are pushed apart. The final result is identical if the electric fields point in the opposite direction ($\rho > 0$ in Ω).*

Independently of the orientation of the spin axis, two particles of the same sign tend to produce regions where the divergence of the electric field ρ is different from zero (see figure 3.2). This is true by virtue of the considerations made in appendix B, based on the observation that the equation $\rho = 0$ cannot survive to sudden dynamical changes of the electric environment. As previously claimed, the stationary part of $\mathbf{E}$ at the boundary of a shell is determined by the inner bare particle (and by the dynamical fields supporting the steady ones). Two equal charges generate conflicting situations in the region of space between the furthermost shells (see figure 3.3). We can for instance localize a neighborhood Ω where $\rho < 0$ or $\rho > 0$ depending on the orientation of $\mathbf{E}$. Note that the product $\rho\mathbf{E}$ always points outside Ω, independently of the sign of the charges (provided they are equal). Since $\rho \neq 0$, equation (87) in appendix G comes into play, with consequences on the gradient of pressure ∇p. I guess that a first variation on ∇p is produced by an amount equal to $-\mu\rho\mathbf{E}$, thus the gradient of p points inside Ω. This says that the pressure inside Ω is higher than the surroundings. The corresponding effect is a pushing on the shells surfaces. Pressure diffuses all around and affects the boundaries of the other existing shells up to the bare particles, that are thrown far apart. In doing this, they also drag the two corresponding systems of encapsulated shells. When the bare particles are at a certain distance two more independent shells may possibly form, but troubles are encountered again when new-coming energy tries to penetrate in between the two shell systems. Therefore, if electromagnetic energy is already present in the neighborhood of particles with charges of the same sign, successive disjoint shells are generated, moving the particles in opposite directions. Thus, pressure waves may travel across the shells. All the shells are then concerned with a momentary "mass" flow. In these transitions, electromagnetic formations, still far from being stable particles, may actually acquire mass and assume for a few instants a state that is in between matter and pure photonic energy. Let me point out that these conjectures could be actually verified by working on the model equations, although with uncommon technical effort.

Let me also recall that, together with the electric field, there is an omnipresent magnetic component. I would like to show that magnetic effects are very relevant in maintaining the equilibrium states of an atom. Neglecting magnetic properties limits the validity of the approach. This crucial observation is also going to be true for Schrödinger's equation and other affine models (see section 3.3). Since $\text{div}\mathbf{B} = 0$, the lines of force must be closed curves. Thus, let me devote some attention for a moment to a few remarks concerning the magnetic properties of atoms. As electromagnetism predicts, these come part and parcel with the electric ones. In fact, although the reasoning is principally based on the electric stationary component, everything is always dynamical and the two concepts of electricity and magnetism cannot be disjoint. We know that an atom reveals magnetic aspects that are often presented in an unconvincing way. At the beginning, quantum mechanics borrowed ideas from classical mechanics, and the concept of magnetic momentum of a particle was inspired by the spinning of a charge. Although not realistic, this explanation is usually the one given in standard textbooks. Later, the idea of spin is introduced in an abstract way, but this further complicates things, since it becomes a kind of dogmatic definition, never clarified and always having an ambiguous link with some unreliable rotating charge. The same was said in section 2.5 regarding the magnetic moment of a nucleus.

Returning to the Hydrogen atom, the spin axis of the electron and the proton can be parallel or anti-parallel. I exclude a priori situations in which the axes are askew, since I attribute higher energy to this eventuality. The problem that emerged a few paragraphs above does not bother us since the electric vectors follow a unique stream. Heuristically, one can expect to arrange things in order to preserve the condition $\rho = 0$, though this is not automatic and must respect the whole set of equations. We can actually find regions where $\rho \neq 0$, due to the fact that the electric signals emerging from the two particles do not match in intensity. Thanks to equation (84) in appendix G, a gradient of pressure may grow proportionally to $-\mu\rho\mathbf{E}$, corresponding, as one may check in simple geometries, to a negative pressure $-\frac{1}{2}\mu|\mathbf{E}|^2$. The two particles end up in attracting each other. The discussion then passes to the magnetic component (we are starting here to appreciate the importance of its existence). An analysis based on the knowledge of an electron's exact solution gives the following relation: $c^2\mathbf{B} = \mathbf{E} \times \mathbf{V}$ (valid for right-handed triplets; see also (104) in appendix G, given for left-handed triplets). This is true for both the statical and dynamical parts independently. Magically, the same equation holds when studying the *spin-orbit* interaction in Bohr's Hydrogen model (see for instance [Eisberg and Resnick(1985)], p.279). If it is true that $c^2\mathbf{B} = \mathbf{E} \times \mathbf{V}$ continues to be valid for the time-dependent fields outside the particles, the coincidence is astonishing since it confirms a close agreement between the physics conceived upon standard considerations (on rotating charged bodies in magnetic fields) and the physics constructed solely on the knowledge of pure fields. Classically, the formula allows us to interpret the rotatory path of the Hydrogen's electron as a consequence of the combined solicitations of the proton's electric field and a suitable magnetic field generated by the rotation of the electron itself (more exactly, by the rotation of the proton, as seen from an observer placed on the electron). My viewpoint is much simpler, the two particles do not move and the existence of **B** is automatically granted by the electromagnetic environment.

I insist on observing that the photon shell system of a proton has little in common with that of an electron. Electron shells do not have a wide range; in fact, they can only provide some fine adjustment. Thus the dominating geometrical role belongs to the proton. There are two main reasons for this dichotomy. First of all, the bare proton is rather complicated and, for this reason, substantially different from the simple electron. Such a divergence gets more remarkable as advanced nuclei are taken into account. Secondly, the magnetic energy present at the boundary of a bare proton is much higher than that of the electron. Let me note that such a property has nothing to do with the measured magnetic dipole moment of a nucleus. In fact, there are situations in which such a dipole moment is zero (see for instance ^{4}He, ^{12}C, ^{16}O), but the presence of a magnetic message cannot be denied (perhaps in the form of quadrupole moment). According to (98) in Appendix G, the density of mass of a particle has been defined to be proportional to $\rho - p/c^2$. Quantitatively, it turns out that pressure p is mainly affected by the contribution of **B**. As the proton is very massive compared to the electron (about 1830 times more), we should then expect intense magnetic fields in its surroundings, transferring high energies to the circulating system of photon shells. From the quantitative viewpoint, the estimated magnitude of **B**, when acting on the dipole moment of the electron in the Hydrogen ground state, is extremely large; here we are talking about intensities of the order of 1 Tesla (see [Eisberg and Resnick(1985)], p.281). The strongest man-made magnets do not exceed 40 Teslas. Therefore, the role of the magnetic component ends up being decisively important.

Allow me to be more specific regarding the fields developing around an electron. The information actually travels within the electromagnetic vacuum, but the intensity is not enough to sustain closed bubbles. There are documented experiments showing that bunches of electrons, traveling at relativistic speeds, may remain focused (see, e.g.: [Humphries(1990)]). This means that the particles do not repel each other as the Coulomb's law would suggest, requiring a mathematical adjustment of the formula. Perhaps, the inherent reason is that, at these regimes, electrons do not find enough strength to communicate their properties of being charged. Instead, they possess the likelihood to easily exchange messages inside molecules via the shells produced by nuclei.

In the atomic structure, the magnetic signals are brought from the respective bare particles, which, as stated before, display different magnetic behaviors. This is also true in the neighborhood of the segment connecting the spin axis, where the magnetic field is zero. Thus, the corresponding intensities may not match, generating a region of strong variation at the interstice. By equation (84) in appendix G, the sudden variation of curl**B** reflects on the term $\rho\mathbf{V}$, that also appears in equation (87) under the form $\mu(\rho\mathbf{V}) \times \mathbf{B}$. At this point, a gradient of pressure ∇p develops, which is of an amount roughly proportional to $-\mu\text{curl}\mathbf{B} \times \mathbf{B}$. In simple displacements, this last term is compatible with a positive pressure equal in first approximation to $\frac{1}{2}\mu|\mathbf{B}|^2$. The reciprocal position of the particles mutates until one has $\nabla p = 0$. The adjustment process aimed at reaching an equilibrium state may pass through the asymmetrical creation or suppression of shells. A perfect equilibrium state could not possibly exist; however, it is reasonable to suppose the presence of a magnetic basin permitting the electron to float. In that location, the resultant force is a combination of both the attracting electric component and a magnetic repulsion that forbids the electron to further proceed in the direction of the proton. This argument applies independently of spin orientation. Both particles induce perpetual oscillations in the electromagnetic vacuum, with nonmatching frequencies, so that their reciprocal assets fluctuate endlessly.

In figure 3.3, it is naively shown how, by varying the distance of the bare particles, one can get an intermediate configuration leading to a smooth joint of the magnetic signals, independently emanated by the two sources. Since the magnetic field of the proton is much more vigorous, the meeting point of the signals must be positioned very near to the electron. If we also recall that the intensity of **B** is somehow related to the mass of a particle, it could be possible to justify the geometrical displacement of the Hydrogen atom from the magnitude of the ratio M/m between proton and electron masses. Note that the parameter M/m takes part in the description of the Bohr's atom through the so called *reduced mass* of the system, and this is another salient coincidence between my approach and classical results. Differences in atomic configurations are noticed for instance by replacing a nucleus with any of its isotopes. It is also known that by replacing an electron of certain atoms by a muon, which is 200 times heavier, the spectrum changes considerably. This confirms that mass actually plays a role in the atomic world, and here we roughly start understanding how the promiscuity between electromagnetism and mechanics may have come about, although nothing is mechanically turning around except photons. It is known that strong external magnetic fields can significantly interfere with the shell distribution providing visible perturbation on the emission spectrum of a substance. The *Zeeman effect* is an example of such modifications, confirming that the magnetic context is latent and ready to come into play as soon as appropriate conditions are fulfilled.

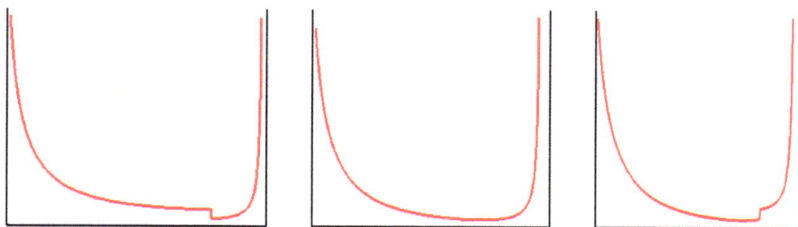

Fig. 3.3 *The signals from two different sources meet at some point in between. It is possible to adjust the distance of the sources in such a way that there is no discontinuity at the meeting point (see central diagram).*

3.2 Photon emission

At the end of the nineteenth century, the scientists J. Balmer and J. Rydberg found, for the Hydrogen atom, that the energy needed to pass from the n-th state to the m-th state is proportional to: $1/n^2 - 1/m^2$. The state corresponding to $n = 1$ is at the minimum energy and it is called the *ground-state*. Jumping from the state $n = 1$ to the state $m = 2$ requires much more energy than moving for instance between the states $n = 10$ and $m = 11$. Thus, the atom can be excited more easily when it is far from the ground-state.

Following Planck and Einstein, the energy of a photon is proportional to its frequency. This is partly true. For a general free-wave (and the photon is one of them), there is no reason (nor possibility) to introduce the concept of frequency, since any kind of information can be carried, varying both in intensity and shape. In fact, when $\mathbf{E} + \mathbf{V} \times \mathbf{B} = 0$ (see (19) in appendix C), the modeling equations do not impose heavy restrictions, so quantization cannot emerge in any form. For photons produced from the breakage of a rotating system, the situation is different. There, we find a preliminary form of quantization that links the spatial magnitude of the wave (i.e., its wave-length) and the inverse of the frequency of rotation, before the rupture occurs. I claimed that, around a proton, there are various photon shells, displaying frequencies decaying with distance. The proton-electron system, together with the corresponding bare particles, may give life, if forced to do so, to further interstitial shells. Passing from an energy state to the next one at higher level, amounts to adding a lower energy shell between particles in a way to be studied later. This can be achieved only if the new-come photons have the right amount of energy to create stable configurations. On the other hand, an excited atom can get rid of the lowest energy shell and pass to a lower state through photon emission.

Without dismantling the work accumulated in more than a hundred years of atomic physics, my set up is in line with the experimental results. The new and decisive instrument for the analysis is a more powerful theory of electromagnetism, which provides a deterministic model for describing photons, allowing for full control of what happens inside Hydrogen. Note that here the word "inside" is very appropriate. In fact, the vision of an atom composed of a material nucleus and a bunch of electrons, exchanging information through not well formalized subparticle carriers, has been now replaced by a thick set of circulating photons. There is no void between the bare proton and the bare electron, but an impetuous rush of structures, animated by different particular frequencies, forming the distinctive imprint of Hydrogen. These photons are ready to come out at any moment, as the appropriate stimulation is applied. From the modeling equations, it is possible to show that high-energy free photons may pass through larger low-frequency shells without being too affected, abandoning the emission site. In some other circumstances, electromagnetic or mechanical energy coming from outside can be encompassed and digested by the atom, then suitably filtered to produce "orbitals" of the right frequency.

In chemistry, orbitals are used to denote the regions where electrons are supposed to be statistically found when an atom is in a certain excitation state. According to my viewpoint, electrons are not orbiting at all and there are no potential orbits where they must be searched for. Orbitals are really existing photons structures, wisely arranged between bare particles. Moreover, the atom, by acting on further electromagnetic energy, can also generate additional outer shells. In this fashion, it will be able to interact with other atoms to form molecules. A kind of action radius can be naturally associated to the proton-electron pair in the Hydrogen atom in its ground state, which is of the order of half an Ångström. It is basically similar to what is called the *Van der Waals* radius, of practical utility in the study of chemical compounds. It is interesting to note that such a domain also follows the contour of the electron. This fact is important since I am going to give a promotion to electrons, which will be handled in a chemical structure in the same guise as nuclei. Hence, electrons come out from anonymity; they are no more fuzzy underdeveloped displacements and they will expect to be treated as respectable components of matter. Having set the basic ideas, let us go now to check some quantitative aspects.

Photon shells do not have charge or mass. In truth, they globally display an energy (at the ground state) given by the Planck hypothesis as the product of the constant h and their corresponding frequency. The total sum of the energies is actually equal to the electrostatic energy of the Coulombian field integrated on the domain of the shells (see appendix I). However, the associated densities of charge and mass are zero. The electric and magnetic properties, expressed by the bare particles, are transferred to the shells. The process is repeated in the successive shells with no influence on the total mass and charge. Together with the electromagnetic information, the shells carry outbound gravitational information. Some ideas on how the formation process of the whole structure takes part will be given in the following paragraphs.

Recall that each shell is an indivisible entity. By the way, a shell can adjust its size and, when it has accumulated enough energy, may start a smooth pre-subdivision process, brusquely terminating with the splitting into separated shells, with a consequent mutation of the topology. In quantum mechanics, the transition between states is explained by attributing to the electron a vibrational movement until it reaches the new equilibrium. During the oscillations, there is a production of electromagnetic waves like in a dipole. Dipole waves are far from being self-contained solitary waves (as a typical photon should be); therefore, this explanation of the photon's emission process remains vague and unconvincing[110]. Another explanation

---

[110]From [Hecht(2002)], p.65: "The radiated light can then be envisioned in a semiclassical way as emitted in a sort of oscillatory directional pulse [...] In a way, the pulse is a semiclassical representation of the manifest wave nature of the photon. But the two are *not* equivalent in all respects: the electromagnetic wavetrain is a classical creation that describes the propagation and spatial distribution of light extremely well; yet its energy is not quantized, and that is an essential characteristic of the photon. So when we consider photon wavetrains, keep in mind that there is more to the idea than just a classical oscillatory pulse of the electromagnetic wave".

usually put forward relies on the observation that an accelerated charge emits a series of photons. Since there is a shifting of the electron from an equilibrium position to another, this explanation seems more credible, though still not coincident with my viewpoint. It has to be clarified however how the mechanism of photon emission during acceleration works and why this is not a continuous process but subject to quantization. The classical arguments, justifying how a moving charge produces electromagnetic field, do not explain what really happens, and, in particular, why the emitted wave has compact support. But, if one agrees with my interpretation, the view gets more clear. Things are however not as simple as just depicted, therefore deeper investigation is necessary. Although I am conscious that details are important in a scientific treatise, the search for a global understanding temporarily takes precedence. So, I ask the reader to forgive me if the dissertation is unavoidably vague.

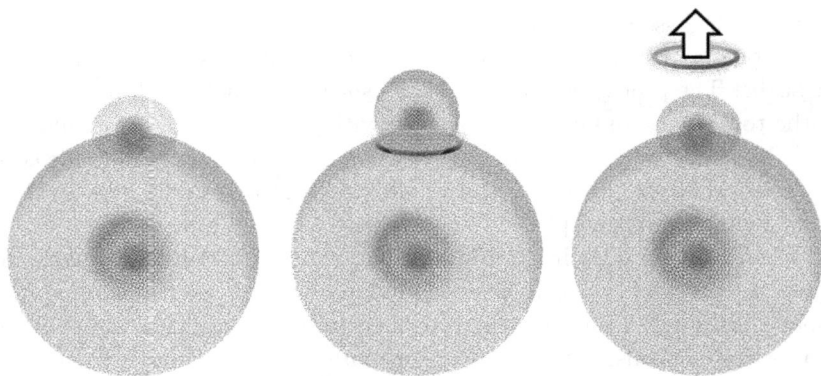

Fig. 3.4 *Approximate sketch of the photon shell distribution in the Hydrogen atom (scales are not realistic). The proton has a fundamental system of shells and the electron finds its position where electric attraction and magnetic repulsion are in equilibrium. Then, after topological mutation, the sequence extends further away. In this simplified picture, spin orientation is not considered. Both particles occupy fixed positions, whereas the effervescence is given by the circulating photons. As the system is perturbed with a sufficient amount of energy, a further disconnected shell, shaped as a ring, may form in between. The structure, recalling that of a scamorza cheese, returns to the ground state after the emission of a photon.*

Figure 3.4 shows what it is supposed to happen during a photon emission in the Hydrogen atom. On the left, the atom is at its ground state.

The particle system is agitated by small mechanical oscillation. Altering more vigorously the structure may lead the atom to "jump" to another state with the creation of one or more shells, taking the form of tiny rings attached to the electron. As the atom returns to rest, the rings are emitted as free-waves. The energy of each ring strongly depends on the context. In principle, there is no prescribed rule for the spectrum of frequencies produced by an accelerated electron and for this reason is said to be continuous. When the electron is tied to a certain atom, it is no longer a single entity but concurs, together with the nucleus and the other electrons (if any), to a quantized formation of shells, specific to that atom and related to its excitation state. This phenomenon is global and it is not directly due to the fact that a specific electron of the group has been accelerated[111]. Changes can be only ascribed to the photon structure since electrons remain blocked in a small neighborhood of their initial position.

Some technical aspects are examined in appendix I. The computations are based on the determination of the field $\hat{\mathbf{B}}$ introduced in the previous section. I recall that $\hat{\mathbf{B}}$ solves a stationary differential problem, obtained from the wave equation for the magnetic field $\mathbf{B}$, combined with the hypothesis that the frequencies of the shells decay proportionally to the inverse of the distance from the source. I also recall that the zeros of $\hat{\mathbf{B}}$ approximately correspond to the separation surfaces of the shells. In the Hydrogen atom, there are two sources and this aspect has to be imprinted in the equation (see (128) and (129)). Moreover, the intensity of the message emanating from the proton is supposed to be order of magnitudes larger than that coming from the electron. This property is obtained by imposing that β_p and β_e in (129) are such that $\beta_p \gg \beta_e$. It is then assumed that the electron is located at a certain distance from the proton, as described in the previous section. The results of appendix I indicate that, when the electron is moved from its position and the perturbation is sufficiently large, a photon ring may originate in proximity, as testified by the fact that $\hat{\mathbf{B}}$ attains new zeros. Such a quantization process descends from observing that (128) is actually an eigenvalue problem in vector form. Boundary conditions are imposed on the particles, having a finite radius denoted by $\delta > 0$. Depending on the mutual positions of the particles, appropriate eigenfunctions are obtained and the set of their zeros may vary following quantum leaps (similar to what

---

[111] From the Historical Introduction in [Born and Wolf(1980)]: "The problem of how light is produced or destroyed in atoms is, however, not exclusively of an optical nature, as it involves equally the mechanics of the atom itself; and the laws of spectral lines reveal not so much the nature of light as the structure of emitting particles".

happens to the number of nodes in a resonant one-dimensional cavity). Of course, the idea can be extended to an arbitrary number of particles (see section 1.3). These computations are however too preliminary. They show a qualitative behavior in agreement with expectations, but as far as the quantitative aspects are concerned, there is the need for a more refined numerical treatment (in order to check for instance that the energies of the emitted photons match those observed in nature).

Let me add more comments concerning inertia. A way to accelerate a charged particle is to apply an electric field. The law is well described in physics books, although such an "action at a distance" is a mysterious process, and, being instantaneous, violates the postulates of special relativity. My approach is different. As we saw, the particle is made of electromagnetic signals confined in a toroidal region. Interaction with an external electric field results in a change of orientation of its internal fields, which also implies a modification at the geometrical level. Geodesics are no longer closed orbits, but start spiraling and, from the outside, one can see the particle shifting. Everything continues to evolve at velocities close to that of light, although the shift can be much slower. At rest, the particle can be compared to a working engine with the gear in the neutral position, waiting for a possible input signal. In some regions, the speed of light may be exceeded. As explained in section 2.2, such a localized phenomenon is not alarming and does not violate general relativity. It is clear that now the interaction is at a purely electromagnetic level and does not need further abstract explanations. An outside inertial observer sees the movement of the particle, which is correspondingly transmitted to the surrounding shells. Due to the larger size and the lower frequency, the motion of the geodesics related to the far away shells occurs with a milder shifting. The shells experience a sort of "inertia", although here this concept is naive since there are no masses involved. At a certain point, some external shells may uncouple from the group and give rise to free photons.

Contrary to what is generally believed, it does not make sense to speak about pure stationary electric (or magnetic) fields. According to my model, an electron generates its own dynamic surroundings, overlapped by stationary fields. The structure however stays in equilibrium because of the time-dependent parts. Thus, everything should be interpreted in terms of shell interactions. In this way, processes at the atomic level are depicted as a movement, destruction and recombination of 3D pieces of a complicated puzzle, with tassels varying in size from the extremely small fragments, directly built around particles, up to the largest intermolecular components[112]. We do not have to actually know what is inside these bubbles, since they are independent units. Although one may reconstruct their internal dynamics through the modeling equations, the observations would be simpler if it was possible to handle them as macro-structures, glued together by surface tensions. The successive theoretical step would then be providing some model equations for the shells, without necessarily computing the exact internal field. The process passes over to geometry and, through it, to gravitation (in the extended sense of the term). We should not confuse this with everyday gravitation, that is only a by-product of a complex mechanism occurring inside molecular structures (see section 2.6).

Due to the oscillating character of the magnetic fields, it is hard to distinguish, in the Hydrogen atom at the ground state, the case when the spin axes of the proton-electron pair are parallel from that in which they are anti-parallel. This difference will be anyway more significant later, when discussing molecules. Nevertheless, as the atom is excited, it is important to know if, at the moment the perturbation occurs, the magnetic fields are parallel or antiparallel, since this actually may result in a slightly different distribution of the atomic spectrum of frequencies. The phenomenon is observed in practice and it is referred to as *fine structure*. Indeed, many emission lines of atomic spectra appear to be the overlapping of two almost coincident bands; these show us the fingerprints of the atoms in both situations (i.e., spin parallel or anti-parallel). The flip of the spin of an electron inside an atomic structure rarely occurs without the intervention of external factors. Spin-flip may be generated for instance by magnetic effects (in the context of quite complex situations however). Changes in the spin orientation are well documented and correspond to the emission of the so called *21 cm-wavelength photon*. It is also known that such an effect is actually due to magnetic reasons. From my viewpoint, the emission of a photon in the transition associated to a spin-flip, amounts to the liberation and the successive breakage of a very low-energy shell. Astonishing similarities can be noticed with regard to electron-positron annihilation (see footnote 63). As a final remark, let me point out that my view of Hydrogen atom is clearly an oriented dipole. This means that, under an application of a congruous electric field, it must rotate and assume an elongated shape,

---

[112]From [Mezey(1993)], p.17: "Motion is an inherent property of molecules; consequently, molecular shapes cannot be described in detail without taking into account the dynamic aspect of the motion of various parts of the molecule relative to one another. Within a semiclassical approximation, the dynamic shape variations during vibrations can be modeled by an infinite family of geometrical arrangements".

with the consequent alteration of the emission spectrum. This is actually known as *Stark effect*. The subject is treated in the framework of quantum theories (see, e.g.: [Rice and Good(1962)]). In my construction, an intrinsic dipole moment would be present even when the atom is at the ground state, a situation not covered by the present theories, although it is proved to be satisfied by some heavy atoms (see [Chupp et al.(2019)]). Deeper investigation is then necessary.

Several concepts are fused in the Hydrogen atom which in the microuniverse represents the easiest example of interacting bodies — the germ of what in a complex and sophisticated context will become classical mechanics. Quantitatively, the electric stationary field is indispensable to defining the constituting structure, and symbolizes what we can directly understand about the object. Nevertheless, stationary components only survive with the support of the dynamical electromagnetic background, whose organization, reflecting quantization, is indirectly observed by stimulating the atom[113]. Both ingredients are essential in the constitution of matter that can be formally separated into a sort of "skeleton" and a "living" part. The first does not exist independently of the second one, but it is necessary to give flavor to the universe (see footnote 155). The symbiosis between quantum and classical mechanics is basically accomplished. From now on, the study only becomes a question of complexity[114] and it will be my duty to show to the reader, with a series of examples, how known facts observed in nature are deducible from this paradigm. Numerical computations performed on the differential model may in the future validate this reasoning and eventually point out possible flaws in my arguments. The important thing is having set up the principal ideas and furnished a partial explanation to the mechanism of formation of atomic structures. In the sequel of this chapter, I will try to see if the whole machinery works by examining and comparing further facts.

In view of discussing more involved atoms and some simple molecules, let me end this section by introducing Helium. Arguing as in the case of Hy-

---

[113]From [Cantore 1969)], p.189: "The conception of atoms as consisting of matter and void is clearly inadequate. If the void should be interpreted in the most obvious sense of the word as nothing, how could this conception give reason for the stability of the atomic aggregate? [...] The atom is clearly a complex, unified entity in which charges and fields interact intimately".

[114]From [Feynman(1985)], p.114: "You might wonder how such simple actions could produce such a complex world. It's because phenomena we see in the world are the result of an enormous intertwining of tremendous numbers of photons exchanges and interferences".

drogen, a sequence of shells, resulting from the convoy of spare electromagnetic energy, may develop around the nucleus (which is now an α-particle). Successively, an electron might be involved in the process of formation of the shells. Presumably, it will find a preferred standpoint along one of the two privileged directions determined by the nuclear proton's axes (see figure 2.14). The electron's spin axis may be suitably oriented in parallel or anti-parallel fashion. Such a combination of electron and nucleus (forming a positive Helium ion) establishes a new peculiar stable configuration of shells, leaving however an option for the arrival of another electron, preferably from the direction opposite to that of the first electron. When the full atom (the nucleus and two electrons) is set up, the situation is saturated and does not allow further additions. Something new however happens, i.e., the two electrons can communicate through the main Helium shell, and I imagine this will bring their spin axes to point in the same direction in order to maintain in resonance their high-frequency oscillating magnetic fields. Technically, one of the spin arrows is directed outward, whereas the other points inward (one spin up and one down). This is the most trivial application of the Pauli exclusion principle. The ionization energies of Helium are 24.6 eV and 54.4 eV. These values show that the couple consisting of the nucleus and one electron cannot be treated as the union of two separate pieces. The distribution resembles that of the negative Hydrogen ion (H^-, one proton and two electrons). Anyway, He and H^- do not coincide, not simply because the nuclei are not the same, but because the whole internal arrangement is structured in an alternative manner. Again, my atom is not just the union of the bare particles but an active changeable entity, filling up an entire portion of the 3D space with an approximate diameter of 10^{-10} meters.

3.3   The structure of matter

Before studying more sophisticated cases, it is necessary to better explain the family of shells encircling a nucleus and their relationship with electrons. For the reasons set forth in section 3.2, such an ensemble of shells is highly energetic, particularly near the bare nucleus, to the point that a reduced factor of 3-4 orders of magnitude is expected when approaching the electrons. The nuclear shells are very few (maybe less than ten) since their growth is geometric. Through mechanisms of the type described in the previous section, the nuclear shells influence the position of the electrons and, once the settle down is completed, are almost immutable. I am going to provide later a deeper analysis of their behavior. For the moment, let us concentrate our attention on what happens near the electrons.

I assume that in the ground state, no further shells are in proximity of the electrons. The electrons can however communicate with the exterior through the shell system of the nucleus. It is wise to recall how adaptable is the electron's shape that, preserving global diameter, may vary its internal frequency by adjusting its thickness (see section 2.1). This is another degree of freedom to be taken into consideration when passing to excited states. Different energy levels, other than the ground state, may be reached via the absorption of photons in the proper frequency domain. In fact, when all the parameters are stretched to their maximum level, it is the moment to change the topological setting through the release or the acquisition of shells. Those close to the electrons remain however very localized, thus they contribute very little to the global organization of the atom they are embedded in. Therefore, there is not a real movement of mass from one configuration to another. My electrons neither revolve around the nucleus nor move when excited. Small adaptations are just allowed. Topological changes only occur in the photonic environment.

I admit that the above claim is in contrast to what is stated in chemistry textbooks. On the other hand, the same texts do not justify why Bohr's theory predicts that the Hydrogen radius grows n^2 times when passing from the ground state ($n = 1$ and a diameter of about 1 Å) to the n-th state, despite the fact that the size of a generic atom does not exceed a few Ångströms. For $n = 4$, one is already exceeding any credible stretching. Of course, one can always say that this reasoning is old-fashioned and already by-passed by modern quantum mechanics, where it turns out that an electron can actually be found far away from the nucleus, although with a very low probability. In this way, some theoretical aspects are safeguarded, leaving however the auditorium with an unpleasant and unsatisfactory sensation. Thus, in the God-plays-with-dice version, having abandoned the idea of a clear picture of what is really happening, quantum theories are able to support an accurate reconstruction of atomic spectra in terms of energy levels. And it is exactly on energies that we also need to orient our study.

The extended optical spectrum goes from the infrared to the ultraviolet. That is the range of all the most basic photon emissions. It is possible to associate a particular energy to the atomic quantized states of a certain atom; however, what we can see from the emission spectrum is the energy difference between states. A frequency is usually attributed to spectral lines, which may also be measured in terms of wave-lengths (in inverse proportion). As anticipated in section 3.1, a free-wave may carry any message with the same support. Therefore, the longitudinal length of a photon is not clearly connected with any specific frequency. There is no periodic behavior pertaining to a photon; it is just a pulse of a certain dimension traveling at the speed of light. The only reasonable way to distinguish photons (not knowing their exact shape) is to measure their energy. Thus, when we see a specific band in the emission spectrum we can say that photons have been released, having an energy equal to the difference in energies between the states before and after emission.

According to my interpretation, the photons emitted by an atom have a width that can be very limited, so that lots of them can be stored in a narrow neighborhood. When an electron gets rid of a shell (figure 3.4), it means that the constituting photon is practically ejected. One can actually observe such a photon, or, more exactly, one may evaluate the exact energy it is transporting. Since there are bands, corresponding to the simultaneous passage from one state to another of several quantized jumps below, it is necessary to assume that the release of shells may occur in groups. Perhaps for reasons due to resonance, the separating barriers between two or more neighboring shells may crack and the signal merges before being emitted. In this way, the associated global photon carries an energy which is the sum of the energy gaps involved. The maximum energy achievable in Hydrogen can be converted to about 911 Å in terms of wave-length in the ultraviolet region. This estimate is outrageously meaningless if one takes into account that the size of Hydrogen is around 1 Å. Clearly, a photon of that extent cannot fit. On the other hand, photons of such energies can be easily accommodated into suitable shells in proximity of the electron.

In truth, the progression from microwaves (that can be really represented through oscillatory phenomena) to optical "waves" crosses a blurred region. Continuous periodic behaviors are imaginable up to certain frequencies, then one enters the domain of quantum physics ruled by discrete photon emissions. The devices used for the detection and the analysis of these phenomena are rather dissimilar[115]. The transition through the various frequency domains of physics is often imagined and described with the help of an ideal turning knob. There are however unfocused transitions be-

---

[115] From [Finkelnburg(1964)], p.47: "... This whole region is subdivided into the region of γ-rays and X-rays, the ultraviolet, the visible, the infrared, the microwave, and the radiowave regions. Each of these ranges requires a specific experimental technique of recording and investigation".

tween microwaves and optics, and between optics and X-rays. The fact that the wave-lengths of the optic region do not agree with atomic extensions is not a problem anyway. Within each range, everything is compatible. Photons emitted by a green source (or another color of the rainbow spectrum) may reach our retina; they excite certain molecules that allow the transfer of the green information to our brain. Atoms emit photons that are successively digested by other atoms. All these processes remain at the same level. The interpretation in terms of frequencies is just a convention to make easier our analysis of facts.

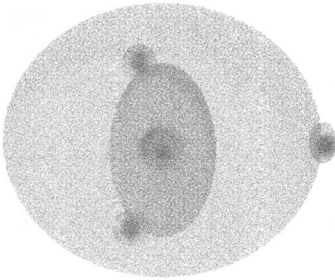

Fig. 3.5 *Qualitative sketch of the system of shells relative to the Lithium atom. Spins are not taken into account. Real dimensions are not respected. Two electrons are at the edge of the K region of the nuclear X-rays systems of shells. The third one is at the edge of the L region.*

 Each electron without wandering around can then eject photons. The process most probably occurs to the peripheral electrons of the atom. The inner ones are instead nailed to their positions which are decided by the nucleus. In Lithium (see figure 3.5), whose nucleus is naively depicted in figure 2.15, two electrons are located near the center and the third one is placed at a distance. In the positive ion Li$^+$, the two electrons are lined up according to nuclear spin orientation. As the third electron tries to complete the neutral atom, it finds no room in the neighborhood of the first couple. An explanation for this "exclusion" should be searched in the displacement of the lines of force of the nuclear magnetic field, following specific symmetric patterns. The presence of the third host is however supposed to interfere with the position of the already existing electrons, that may shift laterally with the aim of achieving electrical equilibrium.

 In classical chemistry language, we would say that the *shell* 1*s* is complete, so that one starts filling the 2*s shell* (these shells are not related to the term "shell" used in my context; I am sorry for the confusion!). In Beryllium, the fourth extra electron completes the 2*s shell*, by settling down in the location opposite to the third one. It is important to remark that, according to my viewpoint, excited electrons just inflate a little so allowing the increase of the number of interstitial photons, but do not leave their site. This is a drastic change with respect to the canonical interpretation. In order to proceed and clarify these sentences, I first need to say something more about the role of nuclear photon shells.

At the beginning, I was very skeptical about the real existence of nuclear photon shells. On one hand, they necessarily result from my theory, on the other hand they are a bit cumbersome. Indeed, they look so peculiar that their detection cannot escape experiment. Then I came across X-rays. These were discovered by W. Röntgen about 120 years ago but their prediction in terms of photon shells was for me a kind of revelation. Wavelengths associated with X-rays are of the order of less than a few Ångströms, therefore they perfectly fit atomic dimensions. This also means that they are highly energetic (intense magnetic fields), as it should be since they live close to the nucleus. X-ray spectroscopy gives very precise information about nuclei. Spectral emission lines in this case are very few but highly representative of each nucleus. This property is in perfect agreement with the existence of my nuclear photon shells. X-rays basically show up when an electron of the atomic structure effectively moves from one position to another, because there is some room nearer to the nucleus not yet occupied. The so called *hard* X-rays correspond to very short wave-lengths below 1 Å, hence in my diagrams they are related to photons circulating between the nucleus and the first group of electrons. They then belong to the so called *K shell* (in X-ray terminology). Softer X-rays are found when moving towards the successive electron zone (see for instance figure 3.5) and belong to the so called *L shell*. In *nuclear magnetic resonance*, intense non-uniform magnetic fields are applied to samples under study. The combined action with a radio frequency pulse allows for the absorption and re-emission of signals that in the *resonant* regime may provide interesting insight regarding the molecular structure of the sample. It should not be difficult to figure out how these phenomena can be incorporated into the framework of my theory.

In light of these new discoveries, we are ready to examine more sophisticated atomic structures. In section 2.5, I gave some ideas for a possible construction of complex nuclei. Whatever is inside a nucleus, I now turn my attention to the qualitative analysis of the behavior at some distance from it. Similar to the case of lighter atoms, I expect the creation of organized photon shells having different shapes depending on the nucleus under consideration. This surrounding energy is extremely important for the characterization of the atom: it is its fingerprint and, as specified above, it belongs to the X-ray range. This has no spherical symmetry, showing instead a certain number of distinguished patterns, peculiar to each atom. An electron may join the system and find a stable spot at a suitable distance from the nucleus (it is available with a set of quantized equilibrium slots). The arrival of the electron changes the configuration of the shells. Thus, we have passed from the system of shells related to the bare nucleus to the one of a positive ion. This modification is substantial, since it affects the position of other electrons joining the group until saturation is reached, which is when the number of electrons equals the atomic number.

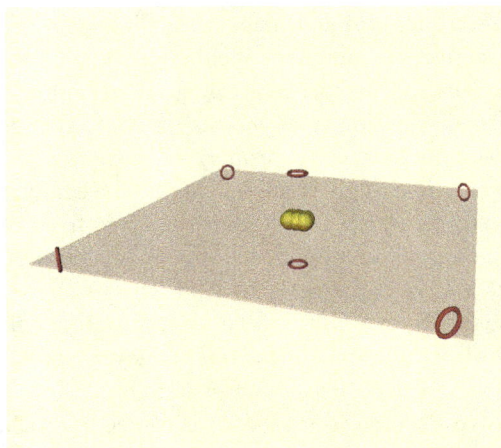

Fig. 3.6 *Schematic 3D representation of the Carbon atom. The position of the electron rings is determined by the photon shells associated with the nucleus. When excited, these photons are emitted in the X-rays range. Other shells may surround electrons and give rise to the optical spectrum.*

Spin orientation of the electrons must also conform. In practice, one should be able to recover all those rules recognized in quantum mechanics by the Pauli exclusion principle. The passage through the various adaptation stages may sometimes be traumatic since photon emission rearranges topological configurations. This construction is not however based on statistics; the position of the involved particles is well defined and their dynamics is due to the rapid circulation (at the speed of light) of the photons taking part and interfering with the outside environment. As in the more elementary Hydrogen atom, the electrons stick on some nuclear shell, maintaining firmly their new position. In the settling down procedure, excess of energy is manifested by the presence of smaller shells for the electron. Finally, all these spurious shells are eliminated, driving the structure to its ground state. In some cases, the intermediate configuration levels may enjoy a large basin of stability so that the ground state may be reached by steps through relatively long periods of time. It is the case for instance in *phosphorescence* and *fluorescence*.

An idea of the electron distribution in Carbon is given in figure 3.6. Nuclei and electrons are not to scale; they have been magnified with the purpose of making the picture readable. In reality, the distance between particles is approximately 10^5 times their size. The configuration is hypothetical, since at this degree of complexity I am unable to be more precise. As the reader can notice, my orbitals disagree with those usually assumed. The $1s$ level is filled, but there is no $2s$ level. The four furthermost electrons are at the same distance from the nucleus. In the Neon atom, I expect eight electrons out of ten to be placed at the vertices of a cube centered at the nucleus. This means that Sodium will have an additional electron somewhere outside such a cube, placed in the transition zone between the M and N X-ray ranges.

	K	L	M	N		K	L	M	N
H	1				Na	2	8	1	
He	2				Mg	2	8	2	
Li	2	1			Al	2	8	3	
Be	2	2			Si	2	8	4	
B	2	3			P	2	8	5	
C	2	4			S	2	8	6	
N	2	5			Cl	2	8	7	
O	2	6			Ar	2	8	8	
F	2	7			K	2	8	8	1
Ne	2	8			Ca	2	8	8	2

Concerning the first 20 atoms, the distribution at the ground state is summarized in the above simplified table, providing the number of electrons occupying positions at the border of contiguous X-ray shells. When the edge of the K level is saturated, electrons accumulate at the edge of the next L shell, and so on. The classification in footnote 116 is amazingly similar. Indeed, it implicitly admits a sort of direct implication between atom's states and X-rays, which in a more recent context are instead not related. It would be interesting to carry out historical research to know more about the reasons for the choice of such terminology. So, according to my view, things do not behave exactly in the standard way, and I think the reader at this point is going to expect explanations.

The energy levels of an atomic structure may be modelled via the *Schrödinger equation*. In the stationary case, the study leads to the computation of eigenfunctions and eigenvalues that respectively represent the states of the quantum system and their corresponding energy strength. In the case of the one-electron atom, the 3D problem can be explicitly solved in a spherical coordinate framework, where the eigenfunctions depend on a certain number of integer parameters. From the mathematical viewpoint the result is a standard evaluation of the spectrum of an *elliptic differential operator*, but for physicists it is an elegant formalization of quantum theories. The first integer parameter is denoted by n and it is directly related to the amount of energy associated with a given state. As a matter of fact, eigenvalues behave as $1/n^2$ in agreement with observations[116]. The two other parameters, l and m, are in relation to the angular momentum and the way the electron's spin is orientated.

When passing from one state to another, not all the transitions are permitted and one has to conform to specific *selection rules*. Moreover, some transitions are amenable to identical energy gaps. Hence, the machinery contains some redundancy[117]. The idea of building weird orbitals to predict the electron's motion is suggested by the shape of the 3D eigenfunctions, therefore it is indissolubly tied to the Schrödinger model. The fact that the equation produces within a reasonable accuracy the energy levels of the different states does not guarantee however that the states themselves are geometrically configured in that way.

---

[116]From [Lothian(1963)], p.122: "Letters are sometimes used to denote the value of n, each letter denoting a shell of electrons (see Pauli exclusion principle). These letters originated with the classification of X-rays spectra. Thus: $\begin{matrix} n=1 & 2 & 3 & 4 & 5 & 6 & 7 & \dots \\ K & L & M & N & O & P & Q & \dots \end{matrix}$ "

[117]From [Eisberg and Resnick(1985)], p.241: "Thus the atom has states with very different behavior, that is with the electron travelling in very different orbits, which nevertheless have the same total energy. Exactly the same phenomenon occurs in planetary motion. This classical degeneracy is comparable to the l degeneracy that arises in the quantum mechanical one-electron atom. The energy of a Bohr-Sommerfeld atom, or of a planetary system, is also independent on the orientation in space of the plane of the orbit. This is comparable to the m_l degeneracy of the quantum mechanical atom".

The organization of the states in terms of the parameters n, l, m, does not emerge in trivial atoms, but may be meaningful in more complex situations (multielectron atoms, molecules, external applied magnetic fields). A good theory is properly fitted when it can handle these generalizations. I am not expert enough to be able to judge to what extent Schrödinger theory can reasonably simulate reality. I know there are difficulties in its implementation. Concerning this, I will add some considerations later in this section. I am here anyway to present my ideas that, being based on purely deterministic arguments, differ radically from the existing ones. Nevertheless, my analysis is not sufficiently developed at the moment to provide an evaluation of the degree of coherence with experimental data. The reader cannot deny that, if my construction was quantitatively correct, it would be superior to standard quantum mechanics in clearness and simplicity.

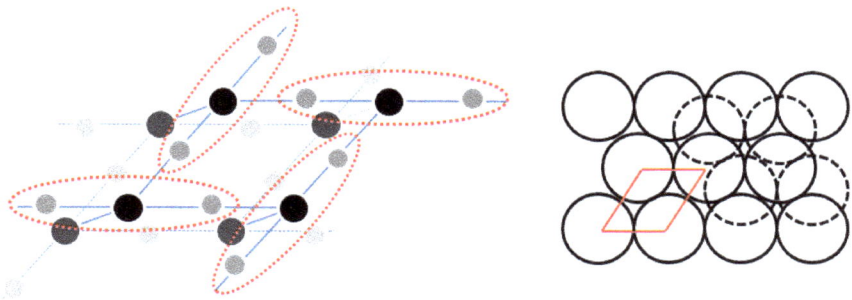

Fig. 3.7 *One of the possible ways Helium may crystallize is in the hexagonal close-packed fashion. To each nucleus (dark larger balls) it is possible to associate two electrons (smaller balls) at prescribed fixed positions. The nuclei are at the vertices of the crystal structure. Dotted lines encircle saturated atoms (one nucleus and two electrons), though there is no unique way to combine the elements of the lattice. The crystal becomes a unitary structure where its parts are homogeneously distributed.*

We can now proceed to the analysis of very simple molecular aggregations. Further external lower frequency shells may develop outside the entire atom which is now ready to form compounds. Helium, being a noble gas, does not take part in the formation of molecules. However, under certain conditions, it displays a crystalline structure. There are several ways this can be done (see [Bailar et al.(1973)], vol. 1, p.167), technically described as: *hcp* (hexagonal close-packed), *bcc* (body-centered cubic) or *fcc* (face-centered cubic). A possible displacement, which is only aimed to show that a combination with stationary electrons actually exists, can be seen in figure 3.7. The larger balls represent Helium nuclei (α-particles). In the "standard" version these are the only visible constituents, since electrons are somewhere in between, distributed according to some probability density. In my version the small balls are the electrons and belong to the lattice. It is not to be forgotten that the main ingredient of the recipe is the complicated system of photons (not displayed here) joining the various parts.

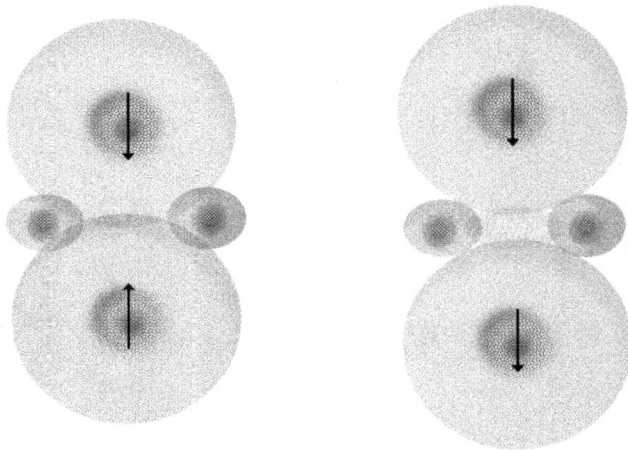

Fig. 3.8   *The Hydrogen molecule* H_2 *in the* para *version (proton spin opposed) and the* ortho *version (proton spin concordant). In the second case, a recirculation ring is present.*

In figure 3.8, we can observe the most trivial compound. In the Hydrogen molecule H_2, one can distinguish among *parahydrogen* and *orthohydrogen*, depending whether the proton spins are anti-parallel or parallel. The first case looks quite natural, having the orientation of rotation of the shells in syntony. In the second case, it is necessary instead to assume the existence of an intermediate recirculating vortex. As a matter of fact, parahydrogen is less energetic and more frequently found at low temperatures. Symbolically, one can write: $H_{para} = H \downarrow + H \uparrow$, $H_{ortho} = H \downarrow + H \downarrow$.

Explaining the reaction:

$$H_{\text{ortho}} + H = (H\downarrow +H\downarrow) + H\uparrow = H\downarrow +(H\downarrow +H\uparrow) = H + H_{\text{para}}$$

is trivial in this framework. More involved is instead the canonical interpretation in terms of electron clouds (see [Rice and Teller(1949)], p.75). A combination with three protons is also known: it is the ion H_3^+ (*trihydrogen cation*). As far as I am concerned, it has the protons at the vertices of an equilateral triangle and the two electrons symmetrically placed on the straight-line orthogonal to the triangle and passing through the barycenter. Other basic molecules are shown in figure 3.9.

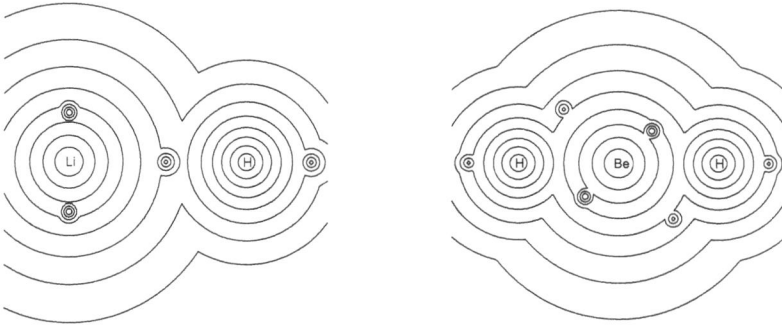

Fig. 3.9 *Nuclei and electron distribution for some elementary compounds. Dimensions and proportions are just qualitative. These are: Lithium hydride (LiH) and Beryllium hydride (BeH₂). The first molecule has four electrons, the second one six. All electrons are suitably located; thus these pictures do not show a distribution of probability but are aimed at indicating exact particle displacement. The qualitative curves encircling the nuclei are related to photons belonging to the X-rays range. Those encircling electrons are associated with photons in the optical range.*

The pictures get more involved as we go to upgraded examples since the number of electrons increases fast. In experiments, an idea of the electron displacement in molecules is obtained by a technique called *neutron scattering* which uses the behavior of neutron projectiles to reconstruct electrical distributions. The results are usually interpreted in terms of a high probability of finding electrons in a determined area, but with my approach, where electrons are blocked in their position, probability becomes certitude.

My diagrams are extremely qualitative and do not give precise indications about bond lengths. A deeper quantitative analysis should be performed in order to check if dimensions are appropriate. At this initial stage, where an accurate analysis lacks even with respect to the simplest configurations, 3D implementations may be challenging. The reader has to confess however that finding a place for electrons in a molecular structure is fun and leads to curious combinatorial problems. So, even if nothing is chemically correct in here, there are at least hints for some new mathematical game.

Starting from the configuration of figure 3.6, four more Hydrogen atoms (each one carrying a proton and an electron) can be added to Carbon. In the first picture of figure 3.10 (top), we may see the final outcome, corresponding to methane. The same tetrahedral configuration characterizes the ammonium ion $[NH_4]^+$. As far as water is concerned (see the bottom picture of figure 3.10), there is only room for two additional Hydrogen atoms since the other free spots have been occupied by electrons. The asymmetrical shape of the molecule is dictated by the form of the nucleus. The three dimensional system of photon shells is not reported since it is rather complex. Regarding NH_3 (ammonia), the situation is more delicate. Here, one of the four free spots is taken by an electron. The asymmetry of the system allows for two configurations that may possibly switch (*Nitrogen inversion*), turning the molecule inside out.

Due to the importance of Carbon, I tried to figure out how basic organic molecules could be made. It is not my intention to rebuild chemistry, but only see if it is possible to distribute electrons in some meaningful fashion. Some of the results of my tests are reported in figure 3.11. I do not exclude that there could be better ways to arrange things in these puzzles. One has to take into account that the single parts have to combine from the electrical viewpoint; so equal charges should stay far apart as much as possible. In addition, each nucleus exerts asymmetric forces, acting like invisible guides and permitting only a finite number of quantized geometries. The exercise may help in understanding the nonlinearity of CH_2 (methylene)[118]. Other cases are displayed in figure 3.12. Let me point out the astonishing similarity of my version of CO_2 with that of figure 6.10 in [Gillespie(1972)].

---

[118]From [Leach(2001)], p.160: "The methylene molecule, CH_2, is of particular historical interest. Despite its small size, this molecule and the controversy surrounding it played an important role in establishing the role of computational quantum mechanical methods in modern-day research and the relationship between theory and experiment. The early debate concentrated on the ground state of the molecule and whether its geometry was linear or bent".

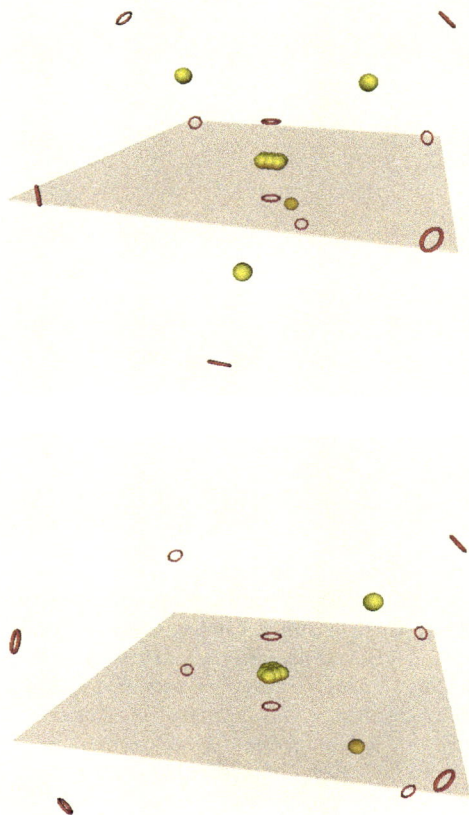

Fig. 3.10 *A possible displacement of nuclei and electrons in methane (CH₄) and water (OH₂). Similarly, one can construct ammonia (NH₃). There are three planes of symmetry in the first case, two in the second case and only one in the case of ammonia.*

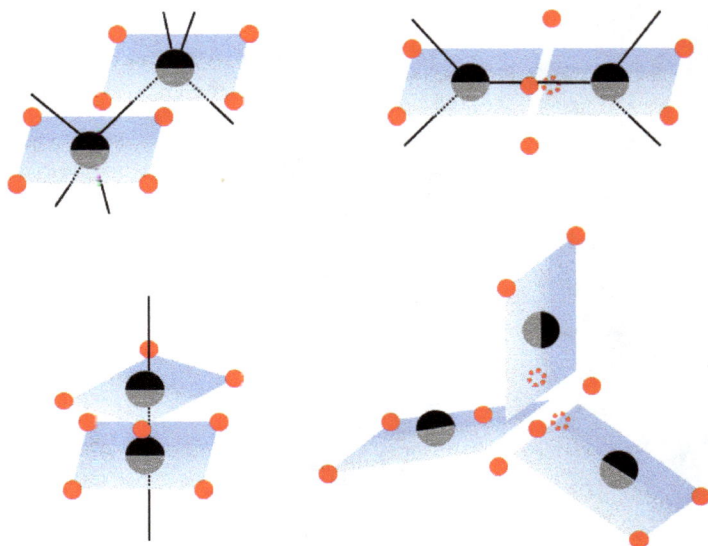

Fig. 3.11 *Possible ways to arrange electrons in organic compounds. According to figure 3.6, in each atom of Carbon two electrons are close to the nucleus but are not shown here. On the top left, a single Carbon-Carbon bond is present, which is the archetype of ethane (C_2H_6). Successively, a stronger bonding is considered; there is less room for other atoms so that one may build ethene (C_2H_4). Note that now four peripheral electrons do not belong to the fictitious horizontal plane of symmetry. On the bottom left, in a triple bond, corresponding for instance to acetylene (C_2H_2), electrons are even more compact. Finally, adjacent are three Carbon atoms: this is the germ of fancier structures of the aromatic type and various allotropes of Carbon.*

Perhaps, the most interesting situation in figure 3.11 is the one associated to the last picture, where three Carbon atoms share part of their electrons. This can be reproduced ad libitum, generating an infinite sequence of composites, based for example on hexagonal stencils. Note that the nuclei are not at the intersection of planes, where instead we find a bunch of electrons. A similar situation is encountered for instance in the O_3 molecule (see figure 3.12).

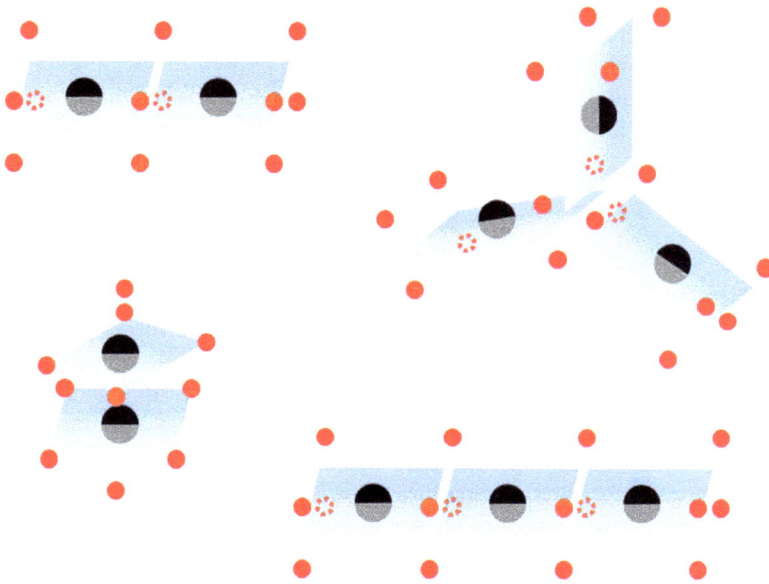

Fig. 3.12 *Electron distribution for some basic molecules:* O_2 *(top left),* O_3 *(top right),* N_2 *(bottom left, see also the triple Carbon bond of figure 3.11),* CO_2 *(bottom right, Carbon is in the middle). Two more electrons (not shown here) are associated with each nucleus. Particulars of the Oxygen atom can be seen in figure 3.10 (bottom).*

The proposed solutions seem to be elegant and put all the atoms at the same level, in contrast with the need of assuming doubled strength for some of the links, in order to preserve the valence of Carbon. On the other hand, in composite structures such as graphene (see figure 3.15), the Carbon-Carbon nuclei distances are homogeneously distributed. These considerations are not new and have been a matter of discussion among chemists[119].

[119]In [Pauling(1960)], p.218, the author writes: "It is true that chemists, after long experience in the use of classical structure theory, have come to talk about, and probably to think about, the carbon-carbon double bond and other structural units of the theory as though they were real. Reflection leads us to recognize, however, that they are not

Benzene (C_6H_6, see also the back cover picture of this book, where electrons are the yellow dots) can be obtained from a single hexagon by placing six Hydrogen atoms close to the intersection of the corresponding planes (figure 3.14). Cyclohexane (C_6H_{12}) may instead accommodate twelve Hydrogen atoms by appending them in dyads directly within the neighborhood of each Carbon nuclei, as done for the lower half of the methane molecule of figure 3.10. The study of anthracene ($C_{14}H_{10}$, see picture aside) starts becoming more complex. The planes of symmetry of Carbon nuclei (related to spin orientations) are here visualized through dashed segments, offering an unusual square-like configuration at the center. Electrons (not shown) are disseminated around nuclei and near plane intersections. Hydrogen nuclei have also been neglected. The present setting may explain the mechanism of cycloaddition of two anthracene molecules, following the absorbtion of UV photons. In this occurrence, the displacement of the global photon system is artificially altered and the central Carbon nuclei find an alternative way to share their electrons, plying the global 3D scaffold and creating the dimer. The reader can easily figure out how to harmonize the various parts.

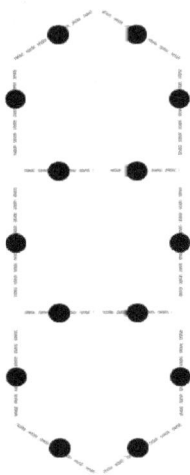

Note that the official explanation of cycloaddition using classical orbitals is quite puzzling (see [Karplus and Porter(1970)], section 6.4.4). Arranging Carbon atoms in order to obtain diamond requires the full three dimensional environment, where planes are perhaps combined to form tetrahedra. Fullerene can be handled more easily. Another challenging area is the chemistry of Boron[120] (the case of B_4H_4 is already hard). I leave this com-

---

real, but are theoretical constructs in the same way as the individual Kekulé structures for benzene. It is not possible to isolate a carbon-carbon bond and to subject it to experimental investigation. There is, indeed, no rigorous definition of the carbon-carbon double bond. We cannot accept, as rigorous definition, the statement that the carbon-carbon double bond is a bond between two carbon atoms that involves four electrons, because there is no experimental method of determining precisely the number of electrons that are involved in the interaction of two carbon atoms in a molecule, and, of course, this interaction has rigorously to be considered as being dependent on the nature of the entire molecule".

[120]From [Bailar et al.(1973)], v.1, p.688: "A unique confluence of circumstances has fortuitously combined to make the inorganic chemistry of boron more diverse and complex than that of any other element in the Periodic Table".

binatorial exercises to suicide volunteers. I give my modest contribution in figure 3.13 with the attempt to visualize hydrazine ($H_2N - NH_2$); by the way, I understand that such an experiment is not relevant without a quantitative analysis (the determination of the bond lengths for instance). The role of Hydrogen in the above compounds may help with a better understanding of the H-bond in more evolved situations (see [Desiraju(2011)]).

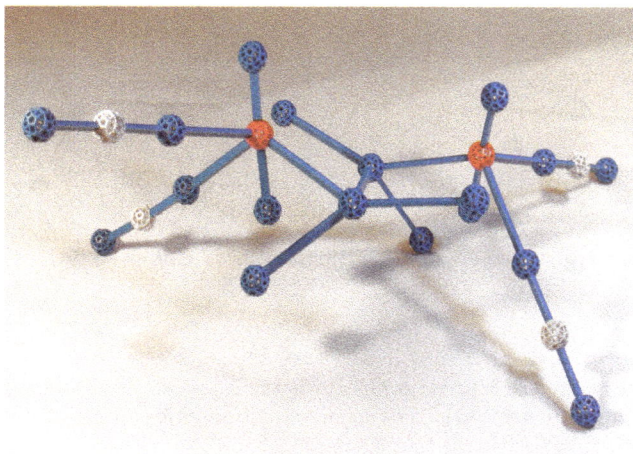

Fig. 3.13 *Tentative sketch of the hydrazine molecule. Two Nitrogen nuclei (red balls), four protons (white balls), and 18 electron (blue balls) are suitably arranged. Sticks are not directly related to chemical bonds. The electrons surround the Nitrogen nuclei, providing at the same time the links between atoms; their uniform distribution is compatible with the bending of the dimer.*

Reproducing the last picture of figure 3.11 to form periodic hexagonal structures leads us to graphene. As specified before in the case of benzene, the novelty here is that Carbon nuclei belong to the segments constituting the net, hence they are not placed on the vertices as is commonly assumed. In atomic force microscopy (AFM) (see, e.g.: [Lorenz and Plieth(1998)], [Capella and Dietler(1999)], [Lipkowski and Ross(1999)]), the surface of a sample is scanned by means of very small cantilever, whose deflection is fine enough to reproduce the geometry at molecular level. Regular patterns as those of graphene are typically visualized (see figure 3.15). The higher intensity (white dots) testifies for the presence of negative charge aggregations. Therefore, positive nuclei, surrounded by the electron cloud seem to be actually localized at the vertices, confirming the quantistic interpretation. On the other hand, in my model, most of the electrons are also concentrated in the neighborhood of the vertices. So, which is the truth?

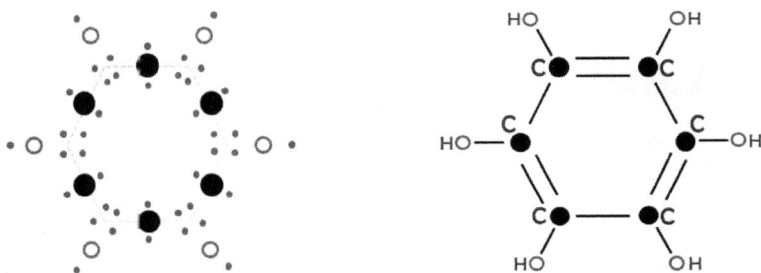

Fig. 3.14   *My version of benzene compared to the classical chemical formula, where the double bonds are arbitrary. The new disposition enjoys a complete symmetry.*

Note that, unlike crystals, many simple organic molecules assume "flat" (planar) configurations (see for instance the sketch of cytosine given in figure 3.19). This is readily explainable by the fact that nuclei are constrained in their relative orientations forming preset patterns and fictitious planes of symmetry (see figures 3.10, 3.11, 3.12). Deviation from flatness is primarily noticed in relation to low-order bonds since single isolated protons (Hydrogen nuclei) impose less restrictions[121]. This noteworthy (and predictive) observation adds strength to my theory. In fact, arguments based on the exclusive use of Coulomb's potential are not enough to justify planar geometries, where additional inputs are needed to make the calculations work[122].

---

[121] From [Leach(2001)], p.457: "The conformations of a molecule are traditionally defined as those arrangements of its atoms in space that can be interconverted purely by rotation about single bonds".

[122] From [Leach(2001)], p.169: "When preparing the input for a quantum mechanics calculation it is usually necessary to specify the atomic numbers of the nuclei present,

My version of cytosine is slightly different from the canonical one, which has well-defined interatomic distances and angles (see, e.g.: [Pauling(1960)], p.307). However, the canonical model is constructed via a comparison of experimental data from varying methods, and final figures result from the application of a mediation process, such as that of molecular *resonance*. Therefore, within certain limits, there should be enough "wiggle room" for different displacements. Assuming one can estimate where electrons are located (as I am trying to do here), a more thorough analysis could be undertaken. The electrostatic potential contour of cytosine is shown in [Leach(2001)], p.84. This has been computed from both the given location of the nuclei and some integral of the so-called "electron density". The outcome is not so dissimilar from that obtainable by grouping the electrons of figure 3.19.

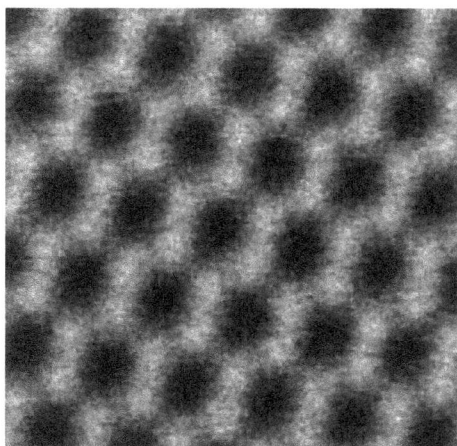

Fig. 3.15 *The stencil of graphene as resulting from atomic force microscopy. Hexagons are clearly represented. By the way, where are the Carbon nuclei?*

Note that photon shells envelop, like onion layers, the single components of a molecule, groups of them and, finally, the global structure. Compounds assume unique photons stratifications that are not deducible straightforwardly from the nature of the single links. Hence a molecule is not the

together with the geometry of the system and the overall charge and spin multiplicity. [...] The atom type is more than just the atomic number of an atom; it usually contains information about its hybridisation state and sometimes the local environment".

sum of its parts, but an entire object with a complex interior and the capability to extend its beingness to the environment[123]. A more extended notion of molecular bond is obtained by involving the concept of van der Waals forces[124]. These promiscuous links are not easily explained by quantum theories, but they can find natural codification in the context of my constructive theory of matter.

Let me devote the last part of this section to some thoughts and concluding remarks. In introducing basic chemistry, nuclei are considered to be sufficient for the description of molecular structures. In fact, electrons are usually neglected. A molecule is generally depicted as a lattice with the different nuclei at the vertices. For example, water is an Oxygen nucleus with two protons at a given distance and a certain angle. This fact is mainly supported by empirical evidence. Fancy 3D graphical software is available to specialists in order to create molecular structures, composed of prepackaged sticks and colored balls already taking into account all the possible bonding possibilities. Actual electrons are totally forgotten and have few chances of directly taking part in the game.

Numerical simulations are indispensable options. Textbooks in physical chemistry show rigorous detailed computations, based on Schrödinger's equation for diatomic molecules, and compare the results with the real ones, magnifying the success of the predictions. They skip however more advanced cases (such as triatomic water), opting for a combination of qualitative theoretical results on eigenfunctions and reliable laboratory measurements. Relativistic effects are sometimes implemented in the Schrödinger model; the reason for this complication being the observation that electrons circulate inside atoms at velocities comparable to that of light (in Hydrogen, the electron's speed is αc, where $\alpha \approx 1/137$). We know that relativity is a key issue in explaining the constitution of atoms; however, according

---

[123]In [Pauling(1960)], p.14, the author writes: "For example, the propane molecule, C_3H_8, has its own structure, which cannot be described exactly in terms of structural elements from other molecules; it is not possible to isolate a portion of the propane molecule, involving parts of two carbon atoms and perhaps two electrons between them, and to say that this portion of the propane molecule is the carbon-carbon single bond, identical with a portion of the ethane molecule".

[124]From [Finkelnburg(1964)], p.376: "... These molecules consist of atoms which are not able to form a genuine molecule by re-arranging their electron shells. However, there exists between them some attraction that is due to interaction forces of the second order. These second-order forces imply the possibility of formation of loosely bound molecules"; and p.416: "Asbestos is an example for a solid in which only fibers or needles are kept together by chemical bonds proper, whereas van der Waals forces are responsible for holding together the molecular chains".

to my interpretation, electrons do not move at all, whereas the dynamical component is given by trapped photons.

The conclusions of existing quantum theories are gratifying, but, together with the limits of the theoretical approach, they set a kind of border beyond which it is not allowed to better model what really happens[125]. Concerns about the criticality and the inadequateness of theoretical quantum theories have been publicly expressed by various authors[126]. A main criticism to Schrödinger equation is that, in its simplest version, it is associated with potential energies built on Coulomb's law. Such a law treats a charged nucleus as a spherical entity and the space around it as an isotropic medium. Therefore, all the directional properties that I am here trying to emphasize get lost. Without additional information taken from experiments, regarding the displacement of the nuclei in a molecule, the plain Schrödinger equation does not automatically furnish reliable predictions[127] (see also footnote 122). Despite the efforts for providing a unifying molecular quantum theory based on the additive union of suitable functional groupings of atoms (see for instance [Bader(1991)], [Bader(1994)]), my understanding, further supported by other evidence that will emerge later, is that electrostatic potentials (and similar other "functionals") are still not sufficient for a global description, whereas magnetism is also required. In conclusion, the process of mediating between theory and experiments is

---

[125] From the introduction in [Pauling(1960)]: "The theory of chemical bond, as presented in this book, is still far from perfect. Most of the principles that have been developed are crude, and only rarely can they be used in making an accurate quantitative prediction. However, they are the best that we have, as yet, and I agree with Poincaré that *it is far better to foresee even without certainty than not to foresee at all*".

[126] From [Maruani(1988)], v.1, p.45 (R.G. Woolley): "Over the last few years there has been a growing awareness that the traditional formulation of quantum chemistry does not exhaust the possibilities for the application of quantum theory to chemical problems. This awareness has come about through a re-examination of the foundations of theoretical chemistry to which many have contributed in the last decade. Ten years ago, I encountered considerable hostility to my suggestion that the programme of conventional quantum chemistry is not just a simple consequence of setting out a molecular quantum theory if one starts from the Schrödinger equation for a system of interacting electrons and nuclei; today that is a much controversial statement, and it is now widely recognized that classical molecular structure is problematic for a quantum theory of molecules".

[127] From [Hendry(2011)], chapter 10: "The Schrödinger equation for the next simplest atom, helium, is not soluble analytically, although accurate numerical methods are available. To solve the Schrödinger equations for more complex atoms, or for any molecule, quantum chemists apply a battery of approximate methods and models. Whether they address the electronic structure of atoms or the structure and bonding of molecules, these approximate models are calibrated by an array of theoretical assumptions many of which are drawn from chemistry itself".

certainly effective, but does not suggest convincing explanations about the foundations of the structure of matter. Hence, the whole superstructure is permeated by an uncomfortable vagueness[128,129].The common trend, especially among physicists, is however to assume that, if a sort of explanation already exists, why look for anything better?

In the effort of mediating between commonly accepted theories of chemical binding and a request of clarity in ambiguous situations[130], in [Linnett(1964)] and [Gillespie(1972)] is analyzed the possibility of localizing electrons in strategic spots of a molecule, following a set of basic rules mainly dictated by the Pauli exclusion principle[131]. The result is very close to what I am trying to do here in these pages. As an example, the NO bound in nitric oxide, where one electron is unpaired in the classical approach, is instead represented with 5 intermediate electrons, with a displacement aimed at reducing the global inter-element repulsion. A similar conclusion (now with steady electrons) would result by suitably combining together my skeletons of Nitrogen and Oxygen.

In my approach, electrons are definitely promoted, becoming active gears of the machinery[132]. A molecule is a lattice of both nuclei and elec-

---

[128]From [Gillespie(1972)]: "Theory has not, however, kept up with experiment, and no comprehensive and completely satisfactory theory for understanding and predicting the structures of molecules has been developed. This is not to imply that we have no understanding of chemical bonding; indeed, several detailed and impressive theories have been developed, but it is none the less true to say that they have not been completely successful in providing a basis for understanding and predictioning why one particular sterochemistry is preferred to another".

[129]From [Maruari(1988)], v.2, p.62 (P.G. Mezey): "It is evident that there is an inherent three-space 'fuzziness' associated with the quantum chemical concept of molecular structure, a fact acknowledged but seldom fully appreciated by chemists".

[130]From [Linnett(1964)], p.155: "In many chemical formulae that appear in books and papers, dotted lines are used to indicate or suggest some form of binding which is less than that of the electron pair. Sometimes the electron distribution or wave function that the dotted line is intended to represent is defined, but this is not always so. It seems to the author to be important and, in fact, to be basic to any treatment that can properly be called scientific, that any symbolism used in a chemical formula should illustrate, in as clear a way as is possible, an electron distribution and a wave function, and that the connection between them should be definable".

[131]From [Gillespie(1972)], p.31: "The purpose of this book is to develop a theory, or more exactly a set of rules, for predicting molecular geometry based on the idea that the arrangement in space of the covalent bonds formed by an atom depends primarily on the arrangement of the electron pairs in the valency shell of the atom which is determined mainly by the operation of the Pauli Exclusion principle".

[132]From [Mezey(1993)], p.4: "In the conventional interpretation of chemistry, the shapes of these fuzzy electron distributions are still much too often relegated to play a role

trons, strongly bonded by a sea of photons. These particles occupy fixed positions, although they can slightly oscillate around in an elastic way like any mechanical device. In this way, it is clear how two or more atoms can "share" the same electrons. The whole system may jump to some excited state by absorbing photons and come back to the ground state by emitting them. Light can strongly interact with molecules to change their chemical properties (see [Flick et al.(2017)]). Due to the distinctive distribution of the photon shells inside a chemical structure, the frequency spectrum of emission characterizes the compound, justifying the amazing results of spectroscopy. In my view, water is an elastic 3D network where we find at its grid-points an Oxygen nucleus, two protons and ten electrons. In the ground state, their mutual positions are exactly determined (up to an intrinsic constitutional tremor), as well as the embedded system of photons shells involved in the links. Similarly, *NaCl* is a 30 body complex. Of course, one can threaten the solidity of the structures by lowering or raising temperature, i.e., by extracting or introducing photons until the bonds are at risk.

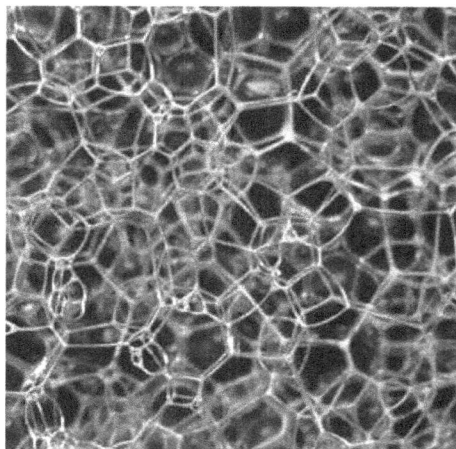

Fig. 3.16 *In solid state, a molecular aggregation can be compared to an ensemble of soap bubbles centered on nuclei. Bubbles also contain electrons, that belong to the interfaces, in order to be shared among atoms. Numerous photons, organized in layers and associated with a multitude of frequencies, are trapped in each domain. They both contribute to stabilizing the structure and carry electric and magnetic information.*

that appears only secondary to the simplest but less revealing skeletons of structural bond formulas. The fact that the peripheral regions of fuzzy bodies of electronic charge densities have a dominant role in molecular interactions is well understood, but it has not yet fully transformed chemical thinking".

For solid matter, heuristic comparisons can be made with a set of soap bubbles joined together (see figure 3.16). Note that nuclei are supposed to be at the center of the bubbles and not in correspondence to their intersections. Such a structure is usually quite stable; if small perturbations are applied the aggregation trembles like a jelly without ruptures. If one bubble of the group is broken by a needle, the others look for an alternative equilibrium state, tending to accomplish the maneuver very fast. Therefore, more ideas may come from the physics of surface tension.

The main drawback of my model is that, even in the simplest case of the Hydrogen atom, full computations get very complicated. On the other hand, if the aim is to incorporate in a unique theory as many situations as possible, all the features must be present and implemented, starting from the foundations. Plots taking into account *molecular electron density* can be found in the literature (see for instance [Gadre and Shirsat(2000)] or [Leach(2001)]). but, differently from my approach, the electron's treatment requires a probabilistic framework. In Appendix I, a stationary differential problem in vector form has been introduced (see also section 3.1). The idea is that the surfaces obtained from the zeros of the solution $\hat{\mathbf{B}}$ may represent a rather good approximation of the shell boundaries. The approach has similarities with that already proposed in [Funaro(2009a)] for the analysis of the Casimir effect. Some rough computations have been carried out for the proton-electron couple in the Hydrogen atom, although better (quantitative) results should be obtained with more serious numerical efforts. The construction has a general validity, thus it can be applied to the study of more complex molecular cases. The advantage is a simplification of the initial set of model equations. At this stage, it is not necessary to know exactly the dynamics of the fields, but only have an idea, as accurate as possible, of the distribution of the shell system for a given displacement of particles (see figure 3.16). It is not my intention to show here any computational result, since the subject is still under investigation. At the same time, one has to look for the right position of the constituents (nuclei and electrons) in order to optimize some sort of functional. Particles must find their equilibrium, which comes from the balance of the electrostatic forces and the constraint imposed by the shells. The construction is highly nonlinear and presents possible discontinuities due to the modification of the topological asset (collapse or birth of a shell). A confirmation that the electrons act nonlinearly comes from observing that the energies needed to ionize an atom assume well prescribed values, that cannot easily be encoded in the context of basic rules. The procedure is made difficult for at least three reasons: numerous individual electrons are taking part in the system; shells surrounding nuclei do not have a simple spherical geometry but reflect the distinctive patterns of each specific nucleus; spin orientation plays an important role especially in conjunction with topological setups. In this way the analysis of a small molecule might turn out to be a nightmare. On the other hand, the aim here is not to speed up computations. The goal is to clarify what is at the origin of molecular organization in terms of a deterministic approach. I am sure that more effective codification algorithms and simplifications can be devised, once a series of studies on basic situations have been carried out. Of course, such a preliminary analysis must confirm the results of practical experiments, otherwise the entire theoretical apparatus is going to collapse.

3.4   In and around matter

So far, we have learned that electrons do not go anywhere. They cannot. They are held firm to keep the molecule in place. At this point, one may argue that electrons in metals usually enjoy a considerable degree of freedom. This property is clearly not true by virtue of a series of experimental observations[133]. Electric signals in copper travel at speeds comparable to that of light (between $c/3$ and $c/2$). In standard conditions, it turns out that the sum of the kinetic energies of electrons is going to be much higher than the energy pumped into the circuit despite conservation rules[134]. Moreover, moving electrons are bare particles interacting with molecular photons. Their passage through a metallic molecule would be like the entrance of a bull in a china shop[135].

---

[133] From [Kittel(1962)], p.242: "... The discrepancy represents an outstanding failure of the classical free electron gas model. It is a particularly puzzling failure in the light of the partial successes of the model in explaining electrical and thermal conductivity in metals: it is difficult to see classically how the electrons can participate in transport processes as if they were free and yet give only a very small contribution to the heat capacity".

[134] From [Feather(1968)], p.335: "When a physical theory has to be buttressed by particular assumptions which do not derive directly from the premises on which the theory is founded, then the theory is failing in a most important respect. [...] The classical free-electron theory of metallic conduction of heat and electricity must be adjudged to be in this category: it provides a pictorial representation of these phenomena which even now is not seriously misleading, but in a wider context it is sterile and unprofitable".

[135] Taken from section 1.1 in [Ziman(1962)]: "If we insist on a particulate, electronic theory of electricity, the high conductivity of metals such as copper or silver is exceedingly difficult to explain. The electrons must penetrate through the closely packed arrays of atoms as though these scarcely existed. It is as if one could play cricket in the jungle".

So, what is really floating when an electric current is flowing inside metal? The answer is simple: photons, the only mobile parts of the system. Zigzagging at the speed of light within the metallic structure, they carry electromagnetic information (and energy) from one site to another. The idea that electric conduction is due to the passage of waves rather than particles has been widely studied, but always within the framework of quantum mechanics, where a point-wise electron is seen as a wave-packet. This brought to the definition of new entities called *phonons*, able to describe transport phenomena in solids. Phonons are vibration waves propagating through the crystal atomic lattice, as a by-product of appropriate compositions of electron waves. They play an important role in judging thermal and electrical conductivity. Packets of phonons can propagate for large distances inside a medium with no significant distortion[136].

Of course this does not match at all with my viewpoint: electrons neither in the form of particles nor in the form of waves travel across molecules. In solids, electrons may enjoy a certain degree of freedom only near the surface; for instance they can be liberated through the photoelectric effect. Studying phonons may give a blurred idea on how energy is transported inside matter. They are supported by an ingenuous implementation of Fourier analysis that however does not provide insight into the heavy non-linear interactions between atoms[137]. There is for sure a molecular agitation able to transfer heat or sound at certain speeds. According to my view, this mechanical information is carried by the deformation (perhaps also accompanied by creation and annihilation) of shells surrounding electrons and nuclei, according to some chain reaction. There are anyway fastest paths where photons can migrate at velocities close to that of light, without bothering too much with low-frequency photon shells, because these latter substructures do not have enough energy to compete with signal carriers.

Evidently, not all materials react in similar ways such that we can make distinctions based on their electrical conductivity. *Dielectric polarization* is due to a rearrangement of charges inside electrical insulators to compensate

---

[136]From [Ziman(1962)], section 1.1: "A single electron is no longer represented as an isolated particle attempting to penetrate the lattice of atoms; out of all the electrons in the solid are constructed waves, which can be gathered into wave packets and guided through the interstices of the crystal almost as if the ions did not exist".

[137]From [Newton(2009)], p.113: "These particles can exist, of course, only inside solid or fluid matter; they have no independent existence in vacuum. Are they 'real'? As far as physicists are concerned, phonons are objects to be dealt with just like photons; the problem of their reality matters little to them".

the action of an applied external field. The phenomenon is permitted by the weak molecular bonds characterizing the medium. The effect is compatible with my interpretation, provided the internal reorganization involves the whole set of shells. What are actually "polarized" are the electric fields carried by the photon shells, perhaps also due to a slight adaptation of the particle spin axes, while the salient points of the molecular lattice in solids remain mechanically firm. The evident promiscuousness of mechanics and electromagnetism inside a piece of matter is manifested by *piezoelectricity*, that is, the ability of some materials to generate an electric field in response to given mechanical strains. Moreover, the piezoelectric effect is sometimes reversible.

Exciton is a magic word that encompasses the capability of the internal structure of matter to change disposition. The transition from a given electronic state to another may be properly described through the quantum absorption of excitons. For instance, this allows the jumping from ground state to the so called *electron-hole state*, which is a way to conceptualize the existence of a positive area suited to host an electron. In some cases, during the return to the atom's bond state, the exciton may be transferred through the molecules; thus, it is regarded as perturbation conducting energy without transporting net electric charge. Excitons are usually studied in the framework of *semiconductors* during the passage from the *valence band* to the *conduction band*. When an electron-hole is created, a nearby electron subject to Coulomb's attraction may take its place, determining the local decay of the exciton. Let me say that this last explanation is more exotic than the one emerging from my conceptualization of matter, where everything is interpreted in terms of evolving quantized bubbles. These may press, squeeze, twist, pass by or disappear, in order to find a state of equilibrium minimizing energy. They leave no "holes" but situations where energy gaps must be filled via electromechanical adaptation. In this scenario, massive nuclei and electrons enjoy a very limited freedom of movement, reducible to a vibrational agitation. In doing this, particles act like pumps relocating the same photon's shells that were responsible for their initial stimulation. Take into account however that electrons are not totally prevented from moving inside a solid piece of matter. In addition to the possibility of vibrating around an equilibrium position, they are free in some materials to look for other compatible stable configurations. This happens for instance in the process of photography, where, under the action of external photons, illuminated areas have their electronic textures reorganized.

Let me now face the issue of *superconductivity*. A way to further weaken low-frequency shells is to subtract energy from a given material by lowering its temperature towards absolute zero. At this stage we know that the bonds between atoms are getting inconsistent and some compounds start manifesting properties of superconductivity. The constituent electrons continue to maintain their positions; if they want to avoid the material becoming a *plasma*, they must guarantee that their shell systems are still at work. Spare propagating photons can now move more easily, basically encountering no resistivity.

Superconductivity is a quantum effect[138]. The accepted quantum explanation is that superconductivity is due to the interaction of pairs of electrons, through the exchange of phonons. The electrons of each pair are not fixed a priori and they can be quite far apart (even ten thousands times the diameter of a single atom of the medium). Their distribution depends on the properties of the conductor crystal lattice. Under the influence of an applied field, each independent pair of electrons linked by a phonon (this system considered as a single particle), travels rigidly and very fast through the conductor, since the way it has been formed exactly fits some resonance wavelengths of the lattice. Again, a good dose of Fourier analysis is the main tool for investigation. Although the quantitative aspects of the theory find excellent agreement with experiments, the explanation looks more an abstract technical exercise than an effective verification of facts.

There is an important side effect, namely the *Meissner effect*, connected to superconductivity, whose explanation in terms of quantum theories is still controversial. Superconductors in the Meissner state are *superdiamagnetic*, meaning that, using a common terminology, at very low temperatures the magnetic flux has been totally "expelled" from inside the body. In other words, at normal temperatures, the lines of force of an external magnetic field pass through the interior of a conductor, while, in the superconduction regime, the external field turns the conductor around, and the inside contribution seems to be completely canceled out. In classical theory, diamagnetism is explained by the tendency of the electrons in orbit around a nucleus to generate a counter-magnetic field when externally perturbed. This should happen by virtue of some momentum preservation arguments.

---

[138] Such a kind of sentence, often encountered in modern physics, looks somehow imperative. Indeed, it does not leave room for other interpretations. It sets a neat distinction among the micro-world and that ruled by the laws of classical physics, as the second were not consequential to the first one. Its use is not inoffensive, denoting in fact a preconceived hostility towards alternative options.

Of course, I disagree with this viewpoint, due to the fact that I claimed that there are no orbiting electrons. At a superconduction level, the Meissner effect is a *quantum effect* (see footnote 138), thus classical reasoning is not accepted.

Whatever the official explanations are, I am going to give my version of the facts. This is quite simple. Photon shells are the carriers of both electric and magnetic signals. At low temperature regimes, the less energetic shells (those associated with low frequencies) almost do not exist, severely weakening the strength of the crystal lattice made of both electrons and nuclei. The electromagnetic field persists inside the higher frequency shells but they look like isolated regions since they are not surrounded by enough energy to communicate their existence. In this way, there is open access without resistivity to new photons, but denial of access to magnetic signals, since the environment does not have the minimal requirements to support them. The external applied field must be very weak, otherwise, since it is also carried by suitable electromagnetic waves, it would raise the global temperature and allow for the formation of stronger links between elements of the crystal lattice, consequently permitting the transfer of the magnetic information inside the body. Experimental observations actually show that there is the creation of a thin internal region in proximity to the body surface (*London penetration depth*), where the magnetic field is not totally canceled. In my opinion, this is due to the slight "heating" actuated by the external magnetic field that sets the molecule's shells, belonging to a surface layer, to an intermediate state of "magnetic conduction".

The existence of magnetic fields inside matter is a natural situation. Officially this is due to the electron spinning, but we already know that this is not the case. There is instead a persistent dynamical magnetic component. In permanent magnets such a component is directly visible; in other situations it is not easily accessible. There are however stimulations that can accentuate its presence. In nuclear magnetic resonance, an applied combination of magnetic fields and electromagnetic pulses to nuclei with spin different from zero allows energy to be stored and then radiated back at specific resonance frequencies. In fact, a constant magnetic field is able to modify the resonant spectrum of the substance, that can be evoked through the action of a specific pulse. Chemical information about molecules can be recovered by analyzing the resonance spectrum at different magnetic field intensities. I would like to emphasize that the frequencies required to achieve resonance are strictly related to the distribution of electrons, which may reduce the effects of the magnetic field at the nucleus (*electronic shielding*). Hence, magnetic resonance is not an exclusive property of the nuclei but of the entire molecular system (see also the previous section). The shielding is also dependent on the orientation of the molecule with respect to the external field. All these observations concord very well with my standpoint.

A piece of solid matter, even in vacuum, is surrounded by electromagnetic radiation. More insight about the presence of this electromagnetic background will be given in the next sections. The internal photons, through their very well organized patterns, give strength and stability to the structure, and extend their action to the exterior. Thus, without the phenomenon being actually visible to us, matter ends up to be embedded in a "dark energy" environment that confers further properties to it. Photons circulate around bodies developing successive stratifications (that I will call *layers*), changing with time, carrying frequencies that decay inversely with the distance. The situation is more complex than that of a single molecule since there are zillions of sources whose effects sum up nonlinearly. There could be however recursive configurations, especially in proximity to the body surface, characterizing the type of material. In this way, shining metal surfaces appear to be different from granular formations. A way to access such information is to examine the spectrum of scattered light. On this I have some remarks to make.

Consulting books and talking with experts, vague and preconceived ideas emerge about light emission, such as for example sunlight. Maintaining the arguments at the intuitive level, there is the tendency to confuse plane waves and photons, a disorientation which is often overcome by stating that the second are the constituents of the first. Even more naively, it is said that photons are the "carrier" of the signal. As I remarked in chapter one, in my theory, the various electromagnetic ingredients are products of a single descriptive model. Unfortunately, from the viewpoint of the narrow framework of Maxwell's equations the explanations remain unfocused and the particulars cannot be appreciated. Plane and dipole waves are usually produced artificially in radio communications or other applications. There are few chances to find examples of this kind in nature. Sunlight (as well as the light emitted by lamps) is the result of photon emission of excited atoms and molecules, through the process detailed in the previous sections. Common light is a multitude of single photons of different size, shape, polarization and "frequency"; an impressive quantity of independent photons marching in the same direction (in first approximation, at least). More attentively, pure free-waves generated by antennas are mathematical abstractions, since a real device, emitting waves through the excitement of an immense quantity of molecules, cannot be such a perfect source.

Recall that we can attribute a frequency to photons, not just because this is in some way written on them, but because they were part of a specific shell of a certain atom, to which we associated a frequency. When light reaches a surface, an extremely complex procedure starts, which cannot trivially be summarized by depicting the parallel fronts of a plane wave while they are reflected or refracted according to the rules of geometrical optics[139]. Note also that plane waves have an infinite extension, thus they turn out to be quite intractable. On the other hand a localization process is forbidden in the Maxwellian case, since it breaks the zero divergence condition for the electric field (see section 1.3). Incoming photons mix up with those already present in matter, transferring part of their energy to the various atom shells. The entire piece of matter elaborates the signals and finally absorbs some of the energy and gives back the remaining in the form of other photons. This is different from claiming that a plane wave (longitudinally modulated by a suitable Fourier expansion in infinite frequencies) hits the body and, depending on the material and the roughness of its surface, part of the frequencies are scattered while others penetrate. The last explanation may work for practical purposes, but it is too rudimentary and does not reflect the complexity of the situation.

Let me finally specify that *lasers* are very similar to sunlight in their constitution, although their photons, enjoying high-directivity properties, display frequencies confined in a very narrow band. Therefore, laser light is far from being a unique focused electromagnetic wave. I know it is useful for engineers to express laser emissions as plane waves, with the fields transversally concentrated in a region with compact support and longitudinally modulated by a certain frequency. I also agree that this is a constructive and effective way to proceed. However, I must warn once again that this is totally inadmissible in Maxwellian theory.

A clear demonstration of the mutualism between atomic organized structures, such as the ones of pure crystals, and the immanence of a photon structure are for example *photonic crystals*. These are periodic

---

[139]From [Feynman(1985)], p.16: "When I talk about the partial reflection of light by glass, I am going to pretend that the light is reflected by only the *surface* of the glass. In reality, a piece of glass is a terrible monster of complexity - huge numbers of electrons are jiggling about. When a photon comes down, it interacts with electrons *throughout* the glass, not just on the surface. The photons and electrons do some kind of dance, the net result of which is the same as if the photon hit only the surface". In my case electrons are almost stationary and the agitation is due to photons.

nanostructures occurring in nature, displaying alternate regions of high and low dielectric constant. Photons may or may not propagate through these crystals, depending on the stimulation frequency. Forbidden wavelengths belong to the so called *photonic band gap*. Photonic crystals are found in many applications regarding the control of the flow of light in suitable optic devices such as high-reflecting mirrors or waveguides. The use of classical Maxwell's equations and plane electromagnetic waves in the numerical simulation of the reflective properties of these materials is a common procedure (sigh!).

In truth, light begins its interaction process with matter long before the arrival at the target. Indeed, matter is a kind of open system; the fact that beyond a certain frontier there are nuclei and electrons does not imply the existence of a real barrier. A system of oscillating energy layers extends outside the limits and covers the crystal lattice structure like an invisible neutral halo. Objects are bigger than expected and they are all connected. At the lowest energy level, the outside photon layers have an energy proportional to the frequency carried. These layers get larger with distance (with geometric growth), hence the energy density decays quite fast, but it can be significantly large near the body to the degree that can severely influence the behavior of incoming photons. External photons of the right energy may not encounter noticeable reactions when passing through the more distant layers, but they may have a hard time if they approach further. I must recall that all these shells and layers are geometrical environments in the sense of general relativity, thus they are associated with significant modifications of space-time. We should not forget that in such regions there are trapped photons that are following curved geodesics of the diameter of the order of nanometers.

The minimum influence that matter may have on an incoming photon amounts to a gentle scattering. This is what is actually observed during diffraction. When light passes through a relatively large aperture, a small diffusive effect is noticed near the rim. We can compare the behavior with that of a fluid subject to a *no slip* condition near the obstacle. The velocity reduction at some points forces the fluid to rotate, producing diffusion (see my numerical simulations in [Funaro(2010)]). At atomic level, the encounter of an incoming photon with the electromagnetic halo around matter modifies the setting of the fields. During this iteration, the photon is no more a free-wave. Accelerations and variations of pressure develop in full respect of energy conservation laws. The photon changes the direction of motion and pressure is exerted on matter, simulating the most simple action-reaction effect. Departing from the impact site, the influence of matter is greatly reduced and the photon returns to the state of free-wave, proceeding straightly at the speed of light. Of course, a remarkable change of trajectory only happens by passing very close to the border of the aperture, while the remaining portion of the hole does not affect the passage of light in a significant way. This is however true provided the opening is not too small. The classical theory suggests that the hole must be much greater than the photon wave length, otherwise things get more complicated.

One of the crucial experiments for the validation of the theory of general relativity was to show that light rays could be deflected by large masses. Einstein predicted that a mass modifies the space-time geometry in such a way that the corresponding geodesics are not straight-lines. Although photons are massless particles, the theory claims that their trajectory may be curved when passing through a gravitational field, and this is what was actually observed by comparing the location of far away stars in two cases, depending on the presence or absence of the Sun in the path between those stars and the Earth. There is no need however to search for astronomical examples to confirm general relativity. The diffusion of light after the passage through a hole is a small-scale experiment showing how strong can be the influence of geometry on photons, causing a deflection of many degrees in a very small portion of space. This also remarks how decisive is the role of gravitation (in the extended sense of the term, as specified so far) in the constitution of matter.

Can we say more about the way photon layers are distributed around a piece of solid matter? Without solving the set of model equations, I can try to guess some qualitative configurations. Suppose that the crystal lattice surface is as flat as possible. Each single nucleus is responsible for the formation of photon shells with decreasing frequency. The operations are almost undisturbed in the vicinity of the nuclei, but their combined interference is felt at a distance. Note also that, due to the nonlinearity of the governing laws, the effects do not simply sum up. In addition, the setting is the result of a dynamical process, therefore I do not expect the regions involved to be immutable. There are a few facts worth mentioning. First of all, regarding size, the formation of nuclear shells follows a geometrical law and a similar behavior is expected for the layers, at least starting from a certain distance. The geometrical spacing of the layers leaves enough room in between for the development of subsystems of intermediate structures due to the influence of neighboring clusters of nuclei. For instance, sitting at a certain point, one can feel both the high frequencies emanating from close nuclei and the low frequencies of the distant ones. Moving a bit spatially, the situation may totally change. Anyway, the transition area, in the vicinity of matter, even if quite troubled, may contain repetitive patterns at different scales of magnitude. It would not be a surprise to recognize fractal patterns there, as a consequence of self-similarity properties (implicitly entailed by the geometric growth).

The literature about chaos and fractals in nature is vast (see for instance [Mandelbrot(1982)], [Peitgen, Jürgens and Saupe(1992)], [Barabási and Stanley(1995)], [Marani et al.(1998)], [Ma, Stoica and Wang(2009)]), so I will not spend further time on the issue. Cauliflowers, sea shells, tree leaves and rock formations, are just a few expressions among the infinite wonders that nature provides us. These are known to be present up to the nanoscale. And then there are clouds, fractures, sedimentations and lightning, just to mention some phenomena that can be more easily associated with time evolution. I am inclined to believe that the indications for the realization of such geometric blossoms are already programmed and hidden in the apparently chaotic electromagnetic halo covering a material surface. The most remarkable example is the build up of a snowflake. When a new water molecule tries to join an already consolidated group, it is not free to approach the system from any direction and angle, but it is suitably driven, in such a way that it may only choose between a few degrees of freedom. When in place, it will give its contribution to deform the global surrounding fractal halo in order to predetermine the possible position of a new entry. Thanks to the information imprinted in the halo, the project is actuated on different fronts that seem to be independent, but are linked by invisible global forces. The final result is a snowflake appearing with its magnificent symmetry. Assuming that water atoms are not conscious of what they are doing, we can hypothesize that their capability to create perfect snowflakes is due to the possibility of communicating through the ubiquitous electromagnetic background. Such communications are not limited to a local level, but are adjusted to pursue a global optimization. This permits the allocation of atoms that are far apart to achieve an unbelievable coordination. Of course, the initial general displacement of water molecules and their formation of tetrahedra are decisive in determining the final shape, but I do not think that the sole analysis of the electrostatic forces, without the help of a global dynamical self-adjustment, would be sufficient to pilot the molecules into such perfect configurations. Take also into account that the mechanism of formation of snowflakes and other organized crystal structures is still to be clarified since, evidently, the real-ization of a minimal electrostatic energy is considered not to be a sufficient argument (see footnote 160). Impressive pictures of wonderful snowflakes are found in the website: *SnowCrystals.com*. Their beauty is emphasized by an astonishing symmetry. It is still unclear however how water finds its way to form complicated stencils, since the sole local electrical interac-tion of single molecules cannot account for a general final displacement of such a regularity. Mathematicians are used to optimizing shapes by work-ing with *variational formulations* based on integrals extended to the whole domain. Once the *functional* to be minimized is given, the analysis turns out to be somehow approachable. The problem here remains at the mod-eling level. If the passage from local to global is not well represented by electric potentials, what is then the right associated functional? I believe the answer should be searched for in the photon shell system. Perhaps, the analysis of the magnetic contribution could be enough. The results of these pages provide hints for understanding snowflakes formation and other similar phenomena.

Let me finally point out that, in the dynamics of chemical processes, topological changes of the photon system, including emission and absorp-tion of new components, are essentially associated to a modification of the magnetic patterns. Under suitable conditions, the magnetic lines of force, organized in closed loops, find alternative dispositions. In this way, an entire closed line can break up into several smaller loops. To mention a concrete example that can be recreated in laboratory, the front collapse of two smoke rings develops into a bunch of little rings escaping transversally from the collision site (see [Lim and Nickels(1992)]). The transition is a splendid geometric metamorphosis within the fluid dynamics framework.

3.5 Quantum properties of matter

By dealing with experiments revealing quantum mechanical manifestations of matter, I am going to enter another slippery issue. Since my fight started with the desire to provide quantum theories with deterministic foundations (see section 3.8), I cannot avoid the duty of facing a more or less detailed discussion. In the previous sections, I empirically analyzed in terms of my deterministic model equations some basic quantum phenomena, trying to describe for instance how atoms and molecules are constituted. I am now going to introduce additional quantum aspects.

I start this section by giving my definition of "vacuum": it is a region of space totally free of electromagnetic radiations. Probably, there are no parts of our known universe where this happens. As a matter of fact, electromagnetic waves are present everywhere in very different forms, although with a common root. To some waves we can associate a frequency, in a way that depends on the mechanism that gave origin to them. For example, waves can be forced to circulate in bounded space regions, or they can be by-products of the breakage of the isolated systems as mentioned above. For objects marching at the speed of light, frequency is inevitably related to their size and energy.

I also claimed that waves carry gravitational information. This is true if we look at them as a geometric phenomenon, essentially related to a modification of the space-time, but practically subjected to the rules of the dynamics of fluids. This is also true because, during wave interactions, pressure wakes may pass by or remain blocked in confined areas, and I suggested connections between the gradient of pressure and gravitational forces (see section 2.6). Such a dynamics is fully described by the modeling equations, presenting electromagnetic waves as a sort of fluid without viscosity. The only difference with a classical fluid is that there is no matter: the properties of the flow are only determined by a velocity field, as it is classically studied in fluid dynamics. We know that material fluids, when suitably driven, can display lots of interesting configurations. Stable structures can then build up also in my context. They have the characteristics of real matter, although they are just made of vector fields[140]. In this way, the link between matter and electromagnetism is soon explained, without assuming postulates based on phenomenological arguments. Everything seems at this point related to the analysis of pure fields, obeying to a set of differential laws that contain the basic ingredients of classical physics[141]. The equations are the result of a wise combination of electromagnetism, fluid dynamics and general relativity, the three main pillars of physics. In addition, they are the prodromes of quantum mechanics. Thus, quantum theories are not directly imprinted in the model, but follow as a result of interacting solutions.

Since I mentioned the word fluid, one might ask if there is any means by

---

[140] From [Innis(1994)], p.121: "When modern physics abandoned the notion of the ether as a kind of 'elastic solid', it arrived at a new concept of matter. 'Matter' became the product of the field, which cannot itself be intuited or represented in and by the imagination. Indeed, there is no schematic *image* of matter".

[141] In [Popper(1982)], p.194, the author claims: "In any case, the fundamental idea of a unified field theory seem to me one that cannot be given up ...".

which it propagates. If there was "something", to be called *ether*, the whole construction of the theory would be quite contorted, since we should start worrying about the presence of other unexpected concepts and entities. There is however no reason to assume the existence of ether; it is in fact enough to assume the definition of vector fields as a primitive idea. So I must take for granted that one knows the meaning of the vectors ($\mathbf{E}$, $\mathbf{B}$, $\mathbf{V}$) and the scalar p. After imposing the initial condition, these objects evolve according to the rules. There are no other elements necessary. I do not intend to investigate what is beyond the essence of vector fields, and I leave the question to philosophers.

I guess that the beginning of our universe started with the release of a huge amount of electromagnetic radiation, corresponding to some initial undetermined conditions. This energy, due to self-interactions, started to create substructures in continuous evolution. Stars are examples of sites where interesting gatherings are forged, but the process involves all the invisible interstitial space, even if matter is not actually present in quantity. In this powerful and immense melting pot, bare electrons and protons are created. I have shown that these can be effectively composed of confined electromagnetic waves, evolving in agreement with the equations and having properties analogous to the real ones. These distinguished tiny packets are the most elementary manifestation of matter, although, other unstable massive structures are constantly generated and destroyed. It is important to remark once again that such a context may be conjectured and understood only with the help of a powerful model for electromagnetism, such as the one I have proposed. Trying to reach a similar goal with old-fashioned techniques is out of the question.

It is evident that matter is just a small part of this immense playground. Even in a small atom, there is a tremendously large amount of energy floating between bare particles. Such an ocean is quite well organized, to the point that it allows the formation of complicated compounds, ranging from the simplest molecules to biological entities. The name *dark matter* is in this case very intuitive, appropriate and trendy, although the term is mainly used in astronomy, where dark matter exists to fill up discrepancies in the masses of galaxies or clusters of galaxies (see, e.g.: [Bertone(2014)]). Cosmologists estimate that dark matter constitutes 80% of the energy in the universe, the rest being occupied by ordinary matter.

My concept of dark matter is however in contrast with what is usually postulated since, according to my theoretical results, it is not possible to assume its existence without a significant background of electromagnetic

radiation. Such a "substance" displays an intermediate state between that
of pure free photons and effective matter. It is a sort of "mushy region".
At the present stage of evolution, the universe is completely filled up with
conflicting waves that leave no room for pure waves. Theoretically, in situations of equilibrium, dark matter is practically invisible. More properly,
we may call it dark energy, since it is made of electromagnetic radiations
in dynamical evolution but with no neat electric, magnetic or gravitational
properties if measured in a certain amount of time much longer than the
various periods of oscillation involved. We can however visualize dark energy. If things are done properly, instruments may detect it, and there are
many ways this can be done. I will discuss this in a while. Indeed, the
abundance of hidden energy is not just the result of my imagination; it is
real and documentable. It is taken into considerations in many disciplines,
although the lack of communication between different branches prevents
researchers from collecting all the snapshots in a unifying album[142].

It is not to be forgotten that shells and layers, apart from their important vibrational contribution, are carriers of stationary electric and magnetic fields. Everybody would agree about the presence of electrostatic
forces, since the interplay between small or large molecules is mostly explained as the result of Coulombian interactions. Less evident is the influence of the magnetic component, which has been proven here to be crucial
in justifying for instance the preparatory rotations and the final corrections
that put the pieces together. In my theory, static fields do not exist alone
but only in conjunction with traveling waves. This makes electric and magnetic fields indissolubly tied, so that, even in the stationary components,
they coexist. The condition $\mathrm{div}\,\mathbf{B} = 0$ puts the magnetic field in a situation
difficult to observe, but its latent presence is ready to come out at any
moment.

A typical weird situation is the following one: magnetic forces are experimentally registered outside an infinitely long cylindrical solenoid (subject
to a current), even if stationary magnetic fields should not be present according to classical arguments. In the framework of quantum theories, the
existence of such magnetic forces is put in relation to the *Aharonov–Bohm*

---

[142]From [Hendry(2011)], chapter 10: "The development of a discipline is the work of
a community of scientists who may be relatively isolated, deliberately or accidentally,
from the work of neighboring disciplines. Each discipline may have its own theoretical
concepts, styles of explanation and judgments of theoretical plausibility, so there can
be no guarantee that physics and chemistry will mesh even if, ultimately, their subject
matter is the same".

effect (see [Aharonov and Bohm(1959)]), which is a quantum effect (see footnote 138). Its discovery was another confirmation of the poor limits of classical electromagnetism, although the current explanations raise other embarrassing questions; one of these considers the possibility that the electromagnetic potentials **A** and Φ (see (7) in Appendix A) contain more information than the effectively observed fields **E** and **B**. Forces are thus insufficient to describe physics, the missing part being attributed to energy potential ([Aharonov and Bohm(1959)] and [Aharonov and Bohm(1961)]). Some claim to be able to directly measure potentials; I instead believe that they remain a mathematical construct with no additional impact on observable events. One can however introduce a novelty in the expression of the potentials in the way roughly described here below.

A solenoid is made of matter and, as far as I am concerned, matter is not the union of its atoms, but something more composite that extends far outside its supposed boundaries through a halo of well organized electromagnetic patterns. The constant magnetic field generated inside the solenoid is a macroscopic consequence, due to the sum of a multiplicity of contributions generated by the electrically stimulated atoms. In the non-convex external part, such a monolithic component is almost absent, since a minimal magnetic part survives in its ground state. In light of this, the arguments justifying the Aharonov-Bohm effect could be reconsidered in the classical fashion, without resorting to bizarre extensions. Indeed, we have observed that there might be pressure jumps every time one crosses a shell boundary. We then combine the scalar p, which is actually another potential (dimensionally different from the electromagnetic ones), in conjunction with **A** and Φ. In this fashion, including both the electromagnetic and the pressure contributions due to jumps, we effectively get more information than that needed to build **E** and **B**. For example, a general Lagrangian function, also incorporating the pressure term, has been proposed in [Funaro(2010)].

Another typical quantum manifestation is the Casimir effect (see again footnote 138). Two parallel uncharged metal plates are subject to an attractive force, even in vacuum (see, e.g.: [Casimir(1948)], [Casimir and Polder(1948)], [Ford(1988)], [Farina(2006)], [Ford(2007)], [Munday and Capasso(2007)]). Such a force is quite strong since it decays as the inverse of the distance of the plates to the fourth power. Due to the presence of small multiplying constants, the phenomenon is normally very mild, but in nanoscale physics the Casimir effect turns out to be quite relevant. There is no classical explanation of this fact and this is an example of a typical problem that can be easily handled by introducing the *zero-point energy*, which is the lowest possible energy assumed by a quantum mechanical system (i.e., the ground state energy). The concept is about one century old and it is used as a synonym for vacuum energy in empty space.

Zero-point radiation pervades the universe and has electromagnetic origin. Such energy remains even when all the usual sources are removed from the system. It is a kind of fuzziness, attributed to matter, at a minimum uncertainty energy level, as a consequence of the Heisenberg principle. Its presence is however obscure in the classical non-quantum context. The theoretical justification of the Casimir effect relies on the fact that the vacuum energy circulating at the exterior of the plates is larger than that trapped in between, resulting in a gradient of pressure acting on the surfaces (see for instance [Milton(2001)] for a technical review).

The influence of the zero-point energy on the constitution of matter was pointed out by W. Nernst long time ago, but the study was not sufficiently developed[143,144]. On the one hand I am glad to hear about these efforts, since a recognized existence of an electromagnetic background is in line with my viewpoint. On the other hand, I consider the modern descriptions to be very simplistic (and vague). In quantum field theory, all the various fundamental fields are quantized at any point of space where a quantum harmonic oscillator is formally placed. Vibrations of the fields then propagate according to a suitable wave equation. In this fashion, one can associate to each point of the vacuum various properties such as energy, spin and polarization. With the exception of energy, they sum to zero on average. The lowest possible energy of a quantum oscillator is proportional to the frequency by a factor $h/2$, where h is the Planck constant. To obtain the vacuum energy in a region of space, one has to integrate over all possible oscillators at all points. Here we start having trouble, since the calculated

---

[143] Concerning the stability of the Hydrogen molecule we find in [Nernst(1916)]: "Die beiden kreisenden Elektronen erhalten durch die Nullpunktsstrahlung ihre geordnete, d. h. im Vergleich zur Wärmebewegung nur sehr kleinen Schwankungen unterworfene Nullpunktenergie, welche einerseits die bekanntlich ser große Stabilität des Wasserstoffmoleküls bedingt, andererseits als mit der Nullpunktsstrahlung im Gleichgewicht befindlich selbstverständlich nicht strahlen kann".

[144] From [de la Peña and Cetto(1966)], p.100: "Nernst also saw that the zeropoint field could help explain atomic stability by providing mechanism to compensate for the energy lost through radiation by the orbiting electrons, and he speculated that this field could well be the source of the quantum properties of matter. Physics went along a different course and Nernst's ideas were soon forgotten; however, we will see them recover their intrinsic value once the zeropoint field is taken seriously into consideration".

energy is infinite (it is usually related to the sum of a series, whose terms grow cubically). Physicists claim that this is not a problem, because they can mathematically handle the situation. I remain skeptical to learn that — although things can be put under control — there is an infinite amount of energy in the neighborhood of any point of the vacuum space. There is of course awareness of the problem and scientists have been trying to do their best to come up with new ideas and interpretations. Efforts to provide constructive models of the vacuum have been made by many authors; some references are: [de la Peña and Cetto(1966)], [Milonni(1993)] and [Ibison and Haisch(1996)].

The attraction of the plates in the Casimir effect is explained by arguing that, due to boundary conditions, only a subset of all the possible frequencies is allowed at the interior, while there are no restrictions outside. The two corresponding energy sums diverge but the first one has less terms with respect to the second one, and this makes the difference. The theory however clarifies neither the nature of these electromagnetic waves nor the way they are actually organized. As the reader already knows, my approach to the problem is definitely deterministic. The electromagnetic radiation floating inside and outside the metal plates follows precise (dynamical) patterns made of a series of layers. In [Funaro(2009a)] I discuss how these non-overlapping films are possibly distributed. The geometry is ruled by an underlying fractal behavior. Assuming that each photon layer has an energy proportional to the carried frequency, the total energy ends up being finite, which is a more sound result. The attractive force turns out to be proportional to the inverse of the fourth power of the distance of the plates, in line with experiments. The analysis is carried out in the stationary case, but the situation is going to be more complex; in reality, the layers, animated by the atoms of the plates, undergo a swarming process of creation, recombination and disintegration. As a rule, high photon frequencies are encountered near the plates. A phenomenon, called the *dynamical Casimir effect* (see, e.g.: [Dodonov and Dodonov(2005)]), has been predicted and deals with the possibility of emitting "real" photons from the system, by suitably accelerating one of the two plates. In the framework of my theory, this is certainly not a surprise. For charged plates with different sign the Casimir effect is enhanced, since more energy is involved, resulting in the stronger Coulomb-type attraction. Finally, when the plates support charges of the same sign, a repulsive phenomenon of hydrodynamical nature is manifested (see figure 3.2).

The heuristic analysis developed so far brings to the conclusion that forces between objects (including gravitational ones), even if they manifest through various typologies, are just different ramifications of a common root. A body reveals its nature by transferring information to an active vacuum and alimenting the surrounding with a distinguishable footprint. Another body can "feel" the arriving message and reacts in several possible ways, depending on its own chemical constitution. A magnet attracts iron and not copper; the interiors of the materials are shaken by the incoming agitation, but in the first case the circulating photons tend to shift asymmetrically, keeping at the same time the chemical skeleton unchanged. Copper is also subject to forces but the resultant is practically zero, or so small to be of the order of standard gravitational effects, as those commonly attributed to masses. Within this framework, every physical phenomenon can be associated to modifications of the curvature of the space-time, with a neat outcome that may be dramatically dependent on the structure and location of the set of shells that fill up the environment. In such a context, gravitation assumes a very general meaning, that may reduce to the mere attraction of massive bodies when most of the main electromagnetic components are in a state of equilibrium with an almost zero average (see also section 2.6).

Based on my scheme, it should not be too difficult to come up with some estimates about the intensity of such a hidden energy inside and around a given molecule. If, quantitatively, the results were optimistic, they may encourage the search for new devices capable of extracting the so-called "energy from nothing" (see, e.g.: [Bearden(2004)], [Cole and Puthoff(1993)]), maybe without the ambitious goal of solving the world's energetic problems, as some practitioners sustain. In truth, most of this energy is the one closing balances in chemical reactions, where the transformation of a compound into another is accompanied by the release of some form of work[145]. The geometrical setting of the reactive substances transmutes into another one at lower energy level, with the corresponding elimination of the exceeding photons. The latter may leave the emission site in the form of electromagnetic radiations (sometimes showing up in the visible spectrum) or contributing to raising temperature. Note that when a complex system is broken in favor of a simpler one, it is hard to come back by recombining the missing pieces, even by replacing the lost energy in some way. This is

---

[145]From [de la Peña and Cetto(1966)], p.198: "The geometry of the energy-extracting device will surely be a matter of careful calculations. Moreover, at such small scales other phenomena probably occur that could prevent the possibility of extracting the energy in a useful way before it is radiated".

because energy is a generic word, while what has been lost in the reaction are the irreplaceable small fragments of a valuable pottery, represented by the initial shell configuration. Such a piece of information cannot be restored and I think this is somehow the justification of the *second principle of thermodynamics*. Suggestions for the construction of "free-energy" motors proliferate in the *web*.

Nevertheless, there is no need to look for fancy applications in order to see how electromagnetic forces can be transmuted into mechanical ones. Repulsion or attraction of charges and magnets, and the functioning of electric motors, are examples in which energies of electrodynamical nature are converted into mechanical work. These devices do not take extra energy from vacuum, but they certainly use the vacuum as a medium. In the context of my modeling equations this is possible due to the mix of electromagnetic terms (the vector fields $\mathbf{E}$ and $\mathbf{B}$) and fluid dynamics terms (the vector field $\mathbf{V}$ and the scalar p). The principles of energy and momentum conservation govern the exchange of information between the various actors. From my viewpoint, a nontrivial problem is to understand how a difference of electric potential can be established between two bodies. For a Volta pile the secret lies in chemical reactions caused by electrodes of different nature. The potential gap propagates along conductors and may "charge", for instance, a capacitor. Since the process physically takes place without moving electrons, the study of this phenomenon is not as simple as it might appear at first glance.

Another area of interest concerns asymmetrical capacitors (see also *Biefeld-Brown* effect). When these devices are charged with different polarities, the two conductors tend to attract each other as expected. The weird thing is that the forces have non-zero resultant, so that the setting is subject to lateral acceleration, and the results, depending on the geometrical configuration, are more or less pronounced. A practical application is the construction of the so-called anti-gravity flying machines (*lifters*). These are very light capacitors, made of aluminum foils, immersed in a dielectric (usually air) and subject to a voltage difference of the order of ten kilovolts. Due to their asymmetry, they start rising and freely floating (see, e.g.: [Bahder and Fazi(2002)], [Canning, Melcher and Winet(2004)]). The explanations of this effect are various, but not really convincing. The most reliable is that there is creation of ions in the dielectric due to a *corona effect*. Thanks to the applied voltage, a migration of ions produces a "wind" that may be better described as a transfer of momentum due to colliding particles. Because of the asymmetric setting, such a wind blows unevenly resulting in a non-zero force on the barycenter. I partly agree with this version. An alternative explanation is that there is no effective movement of ions. What flows are not necessarily massive particles, but interstitial photons (the ones constituting, together with bare particles, the structure of the dielectric, or more generally the background radiation). Forces are generated because such photons acquire a momentary "mass", rapidly converted into work. The reason is due to both the asymmetry of the capacitor and the presence of the dielectric, that in combination give a charge density $\rho = \mathrm{div}\mathbf{E}$ slightly different from zero. Having $\rho \neq 0$, thanks to the model equations, a gradient of pressure may form, with a resultant impact on the mechanical behavior of the entire device.

As an alternative to the idea of the photonic wind, I also suggest the following argument, which can be applied in absence of interposed dielectrics. Near the surface of the charged conductors, at a distance comparable to the atomic scale (10^{-10}–10^{-9} meters), ρ is subject to a very sharp variation: it has a neat value $\rho \neq 0$ within the metal lattice and drops to zero in the relatively extended region between the conductors. Certainly ρ is not zero inside electrons and protons; however, at the nanoscale level, one can assume the existence of intermediate areas (dynamically changing in time) where ρ is also not zero. This is due to the everlasting vibrations of the molecular photon shells, combining electrical and mechanical stresses, as in a micro version of the piezoelectric effect. At macroscopic level, since we are in the electrostatic case, $\mathbf{E}$ is orthogonal to the conductor's surface, so that ρ corresponds to a normal derivative. Let us take $\mathbf{V} = 0$ and $\mathbf{B} = 0$ in the model equations, so we obtain: $\mu\rho\mathbf{E} = -\nabla p$ (see also (109) in Appendix H). Hence, by projecting along the normal direction to the surface, one finds out that a negative pressure $p = -\frac{1}{2}\mu|\mathbf{E}|^2$ is present near the conductors. Given the size of one of the two capacitor plates, the intensity of $\mathbf{E}$ in the neighborhood of the other plate is inversely proportional to its area. On the other hand, as seen above, pressure behaves as the square of the intensity of the electric field. Thus, the two plates ensemble should feel the action of forces with resultant different from zero, that are continuously supplied through the imposed difference of potential. I am afraid however that these forces are extremely weak to be responsible for sensible accelerations, as confirmed by tests conducted in vacuum, where asymmetric forces are absent or too small to be of interest in applications. Recent experiments regarding electromagnetic radiations trapped in asymmetric cavities show the possibility of generating very small pressures on the walls, so producing thrust in violation of the principles of momentum conservation (see [Brady et al.(2014)]). These effects are still the subject of investigation.

The fluctuations of a permanent magnet above a superconductor (*magnetic levitation*) could be partially explained as follows. The closed lines of force of the magnet are prevented from entering the superconductor, since the background radiation in and around it is too weak to support magnetic fields (the phenomenon is known as the Meissner effect; see previous section). The loops are then deviated in conjunction with a general reorganization of the layer system between the two bodies. As a consequence, pressure waves develop, establishing equilibrium situations (*quantum locking*). As the above justification is very qualitative, and the potential p is activated when $\rho \neq 0$, we require a deeper motivation that must rely on the behavior of the electric field.

Theoretically, a source of gravitational field (or a shielding of the Earth gravitational field) could be realized with the help of electromagnetic signals by isolating a significant portion of space where $\rho \neq 0$. According to equations (87) and (98) in Appendix G this would possibly imply $p \neq 0$ and $\rho_m \neq 0$ (note for example that photons actually have $\rho \neq 0$ at their interior and that one has $p \neq 0$ when they collide). Unfortunately, checking this prediction is hard because of the very small constants involved (μ in particular) making the effect practically undetectable[146]. Achieving this goal, even in very mild form, would be a decisive validation of my model equations, and a substantial step ahead towards a new physics.

Let me briefly mention another question that is related to the existence of a non-trivial vacuum (more organized than that predicted by quantum theories). In the *Unruh effect*, two observers moving relative to each other in accelerated fashion have different perceptions of vacuum. This should be true whether we describe the zero-point radiation in the quantum way (a system of connected point-wise resonators) or with my approach. The underlying argument, based on general relativity, is the impossibility in certain situations of finding a suitable mapping relating the observers' frames. I am not going to discuss any theoretical detail about this phenomenon. It is sufficient to point out that the Unruh effect may cause the decay rate of accelerated particles to differ from inertial particles, and this could justify

---

[146]From [Boyer(1975)]: "However, the situation is seen to be vastly more complicated when one realizes that random electrodynamics is intended as a theory of atomic structure. Thus as outlined below, zero-point radiation is expected to provide the random forces which prevent electrons from falling into nuclei. One does not observe the zero-point radiation which maintains the structure of the molecules in the eye or of a mechanical detector, but only the radiation above the zero-point background".

for instance the anomalous behavior of kaons (see section 2.5). In other words, the properties of vacuum may effectively affect the history of a certain phenomenology, as reported by experiments. Further disquisitions on the Unruh effect lead to reflections on anti-gravity and the meaning of absolute time. I leave the discussion to the experts. There is however a *fil rouge* connecting all the above described experiments. The strong isolation of the various disciplines from the general context amplifies the lack of dialogue[147]. The viewpoint I am presenting here may help to unravel the problem and detect the correct links.

The above disquisitions about some well-known quantum properties of matter give emphasis to my unifying approach. Scattered and unconnected explanations now have a place in a more organic and universal context. As I continue to stress from the very beginning, two ingredients are the key: the necessity of a review of electromagnetic theories and the existence of an organized background radiation. I can now deal with more tough and intriguing phenomena. There are in fact experiments that reveal more involved quantum behaviors. These are the starting point of upgraded disputable theories that are the subject of the current research in theoretical physics. Though serious, they open a Pandora's box to sometimes extravagant and "at the limit of credibility" descriptions of our universe.

3.6   More on quantum properties of matter

The *double-slit experiment* is a decisive turning point in quantum physics. Electrons (or other particles) are sent against a barrier with two thin vertical slits at small separation distance (here small is relative to the wavelengths involved). With 50% probability each electron passes through one hole or the other. What can be seen on a screen put on the other side of the barrier is amazing. Instead of two neat marks, a set of interference patterns (bright and dark alternating bands) is observed, "confirming" the wave nature of particles. Indeed, if we accept as given the wave-particle duality dogma, such an experiment is not extraordinary. A trivial explanation comes from the study of undulating phenomena, where at the passage through contiguous holes, the same interference waves are observed. The De Broglie type waves assigned to each electron give rise to interference similar to that produced by waves on the surface of a liquid. To better understand the comparison, one has to assume that such a flow of electrons

---

[147] A Murphy's law says: "An expert is one who knows more and more about less and less until he knows absolutely everything about nothing".

happens in large numbers. Let us admit for a moment that this is the simplest way to provide for an explanation without invoking more involved mechanisms, although one has to pass through the unappealing description of a particle as a peregrinating object carrying in its trip a rather extended wave.

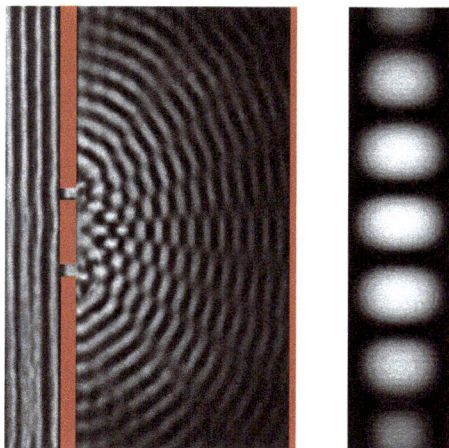

Fig. 3.17 *In the famous double slit experiment, particles emerge from the apertures and interact, forming precise interference patterns. Here, I support the idea that the patterns are pre-constituted; they are byproduct of the footprints left by the apparatus in the electromagnetic background. The particles only serve to illuminate a situation that already exists.*

But the story does not end here. The same patterns can be observed when many electrons are sent one at a time towards the target. Again, each electron has a 50% probability of passing through one of the two slits. When it reaches the screen a mark is left. At the end of the test, when a large number of particles has been shot, the union of the marks does not form two well distinct clusters, but appears with the same interference fringes as the case in which electrons travel altogether. Providing an explanation to this situation is far from being trivial. We cannot argue that electrons interfere with each other, since they move alone, with no apparent way to communicate. In the most favorite version, each particle passes through both slits at the same time (the magic of quantum mechanics!), but other variants (as for instance the *pilot wave* theory), try to explain the phenomenon with arguments more akin to the classical ones, but still far from common sense.

The above example, and many others of the same kind, is a puzzle that opened new frontiers in physics, accompanied by disorientation at philosophical level. Something, displaying a wavelike nature, goes simultaneously through both slits, interfering with itself despite the fact that there is only one particle present at a time[148],[149]. The question forced a review of the most commonplace concepts in physics, including for example the possibility of predicting the future (a particle knows that another one will come and starts interacting with its ghostly version) or assuming a possible correlation of the involved objects (the particles of the experiment are somehow *entangled* even if they occupy different positions in time and space). Quantum entanglement implies for instance that when the first of two correlated particles is measured, the state of the other is automatically known at the same time without measurement. This is true regardless of the separation of the two particles which can be very far apart. As a consequence, there is a kind of information which is able to travel at speeds well beyond the velocity of light. The so called *hidden-variables theories* have been introduced in order to account for the creation of entangled ensembles. Hidden variables permit entangled particles to communicate independently of their distance, violating the basic principles of Einsteinian relativity. For this reason Einstein himself was very doubtful concerning this version of facts.

I would like to avoid spending more words on the above theories and other alternatives. First of all, not being an expert in the field, I risk venturing into vague dissertations that could be easily argued against by a careful reader. Secondly, I dislike the quantum mechanics approach (I am not the only one for sure), since it washes away our certainties, offering an objectionable view of our universe based on elements that are remotely sound but quite disconnected from our common perception. We are at the point that some researchers even believe that our mental attitude is not suited to understanding "matter"[150]. Last, but not least, I have the

---

[148] From [Dirac(1967)], p.9: "The new theory, which connects the wave function with probabilities for one photon gets over the difficulty by making each photon go partly into each of the two components. Each photon then interferes only with itself. Interference between two different photons never occurs".

[149] From [Nikolić 2007)]: "The main reason is the fact that the standard interpretation of quantum theory does not offer a clear 'canonical' ontological picture of the actual processes in nature, but only provides the probabilities for the final results of measurement outcomes. In the absence of such a 'canonical' picture, physicists take the liberty to introduce various auxiliary intuitive pictures that sometimes help them think about otherwise abstract quantum formalism. Such auxiliary pictures, by themselves, are not a sin. However, a potential problem occurs when one forgets why such a picture has been introduced in the first place and starts to think on it too literally".

[150] From [Cantor(1969)], p.287: "Finally, the classical mentality had exaggerated the

chance here to present an explanation in line with my understanding of the structure of matter. The following discussion will be remarkably audacious and I have no idea if it actually matches reality. The hope is to provide fertile ground for possible new developments.

I start with a clarifying remark. The earlier version of the double-slit experiment uses photons as particles. In this case there is no ambiguous wave-particle argument, since one is dealing with pure electromagnetic emanations. Therefore, the interference behavior can be somehow predicted. As I already explained in chapter one, there is a sort of blurred area when discussing these issues, and one never knows when clearly the notion of a wave (as a solution of Maxwell's equations) ends and that of photon, the carrier of electromagnetic radiation, begins. In the end, we always return to the same dilemma: what exactly are photons? By adopting my model equations, all problems disappear. My equations clarify that a solid piece of electromagnetic wave and a myriad of independent photons marching along in the same direction are distinct phenomena, although they are events of an identical nature. An oscillating dipole and a light bulb both emit electromagnetic waves, but with very different characterizing properties. From the discussion in section 3.4, we also know that a laser is a photon emitter and its beam is not a continuous fluid, but an oriented flux of scattered pinheads. Laser photons are isolated waves that are not conscious of being part of an organized controlled emission. Therefore, the argument that photons are waves (not necessarily of electromagnetic flavor) and produce interference patterns (as waves are expected to do), cannot be trivially utilized to give a meaning to the classical double-slit experiment. My photons are strictly localized electromagnetic waves that do not carry any other wave of different typology.

Let us suppose, as it is usually done, that the whole experimental apparatus is made of mathematical entities: the barrier is a geometrical plane, with two perfect vertical cuts. Photons passing strictly within the interior of each hole just go straight through, while those that partly encounter the boundary of the holes may be affected, like solid balls, by some deviation that one may call diffraction. With this setting it is hard to justify the

---

explanatory power of physics. This science was thought to be able to give an absolute explanation of reality by means of an unending series of mathematical deductions based on an exhaustive intuition of the essence of matter. Quantum physics has clarified the meaning of physical explanation. This does not go beyond the ability of tracing back the properties of the compounds to the observed properties of the components. The reason is that human mind is not capable of intuiting material essence at all".

formation of interference patterns. Some may start thinking that my viewpoint is wrong, since the interference bands are actually visible in reality.

But in reality the barrier is not an aseptic geometrical obstacle, but a true and complex piece of matter, and it behaves as matter. We learned that matter is a complicated bustle of photon shells and layers, extending beyond the limits of its molecular set. This superstructure involves a nontrivial modification of the space-time, according to the axioms of general relativity, into independent and evolving elementary cells (see for instance the bubbles of figure 3.16). There are photons trapped in these local geometric environments. The space-time alteration is so strong that it keeps these photons blocked and following extremely curved geodesics. Thus, the whole experimental setting is immersed in an invisible turbulent electromagnetic cloud, exhibiting an extended range of energies and frequencies.

A new coming photon is certainly affected by such a background, especially in the areas ruled by high frequencies. The passage of a solitary photon might also influence the setting, but I expect this happens with less strength, especially when moving at relatively small distances from atomic nuclei. Hence, the path of the photon is deviated and deflected many times before reaching the screen target. The scattering angles are however too small for direct detection. The travel trajectory heavily depends on initial conditions; a small modification of the photon's entry displacement may lead to a drastically new evolution history. This is also due to the fractal nature of the electromagnetic background surrounding matter. The photon is interfering, but it is not doing this with other photons of the family it belongs to (both in the case they are all fired together or one at a time), but with the tortuous environment pertaining to the experiment's site[151]. The same can be said for other kinds of particles.

In practice, the incoming particles of the double-slit experiment only illuminate a preexisting situation. Hidden to our eyes, there are already interference patterns extending between the openings of the barrier and the arrival screen. It is necessary to assume that such an imprinting of the environment does mainly depend on geometrical considerations and very

---

[151] From [de la Peña and Cetto(1966)], p.103: "Any nearby body modifies the background field, and one may conceive that in the neighborhood of a periodic structure the radiation is enhanced in the direction of the Bragg angles. Under the assumption that the electron responds mainly to the waves of the zeropoint field of wavelength close to the de Broglie wavelength, their main effect on the particle will be to produce those angular deviations that tend to give shape to the observed interference patterns. Hence the particle needs to 'know' nothing: it is the random background that carries the required 'knowledge' and operates accordingly to the particle".

little on the chemical composition of the materials involved. If there were no holes, we should encounter an electromagnetic distribution of photon layers similar to that described in [Funaro(2009a)], introduced to study the Casimir effect. If we drill a single slit, the situation is altered, allowing for a new design of the environment, pre-arranged to handle diffraction of incoming particles passing through the aperture. When we drill another slit at a short distance, the interference of the various photon layers prepare the ground for the final configuration. The experiment is done before we actually perform it. Sending particles into the system shows us what really has been prepared down there. Heuristically, I expect the major deviations of the trajectories to happen when incoming photons hit the separation surfaces between the various shells and layers forming and surrounding matter. This should be similar to what happens in refraction, during the passage of light rays through the transition zone separating media of different refractive indices. The final arrival points of these piece-wise independent paths are not uniformly distributed, reflecting instead the prearranged organization of the entire experimental setting, which also includes the arrival screen and the measuring instruments.

Fig. 3.18 *The electromagnetic environment displays organized patterns around the double slit site. The displacement may lead to the impression that crossing particles behave as if they were carrying some kind of wave. The wiggles are instead already present.*

I tried a rough experiment by supposing that each material point involved in the double slit experiment produces a set of shells ruled by the relation $(1/\sqrt{r})\sin(\zeta \log \frac{r}{\delta})$ (similar to the function (122) proposed in appendix I), where $\zeta > 0$ and $\delta > 0$ are suitable parameters, and r denotes the distance from the sources. For simplicity, the contributions have been summed up linearly, though this is surely incorrect. In this fashion, I got the sketch of figure 3.18, where the sizes of the various layers grow geometrically. An incoming particle should feel the presence of such a modified environment in a way that still needs clarification. What happens is probably very complex; there are however elements for a deeper analysis.

If the explanation given above may be considered too involved in the case of a swarm of incoming particles, it is certainly adequate in the case of solitary particles sent one at a time, without the need of assuming exchange of information between entangled particles through other space dimensions or at speeds possibly larger than that of light. The astounding discovery (following the more basic double-slit experiment) that particles independently shot behave identically to those marching all together, should have suggested reconsidering the whole of quantum theory from the very beginning. This viewpoint has been stressed with emphasis[152] within the context of the theory of *elementary waves* (see [Little(1996)] and [Little(2009)]). In short, matter emits elementary waves that interfere non-linearly. Reconsidering all the basic and advanced experiments in quantum mechanics, the study of elementary waves allows for a reasonable restatement of quantum physics without denying evidence and putting things in a unifying and functioning framework. On a similar track is the theory of *Stochastic Electrodynamics* (see [de la Peña and Cetto(1966)]).

Another crucial aspect is the role of a material observer placed in the experiment's site to report on what is going on. It is noticed that the presence of the observer heavily affects the outcome, disrupting for instance the interference lines. Thus, trying to better understand the secrets of quantum phenomena with practical intervention may substantially alter the scene. Such an observation led some physicists to theorize that the sole act of watching makes events happen or not. At a philosophical level, somebody was also induced to conjecture that real things could not even exist without observing them. From my viewpoint the situation is clear: the device used to record observations modifies the experiment because it is made

---

[152]From the introduction in [Little(2009)]: "Some 80 or 90 years ago, physicists made a fundamental error in their development of the theory known as quantum mechanics, the bedrock theory of modern subatomic physics. Because the theory is erroneous, physicists inevitably began to uncover laboratory evidence that contradicted it. In the face of that evidence, physicists should have retraced their steps until they discovered the error; but instead, reluctant to give up the partial success they had achieved with the theory, they chose to 'twist' reality in an attempt to make it agree with the theory".

of matter. When approaching the experiment, it changes the background topology before the experiment is done[153,154]. As a matter of fact, more recent versions of the double-slit experiment, with a less influential role of the observer, partially maintain the features of the original.

On similar lines is the *EPR* (Einstein-Podolsky-Rosen) paradox. I will briefly mention it, with no claim of being exhaustive. In the easiest version, particles from a common source are emitted in such a way that they march in opposite directions. Two measuring instruments are set far apart in order to compare the spin orientations of the two independent particles. Without adding further details, it turns out that the measurment outcomes give an impression of a sort of information exchanged by the particles, that goes beyond basic notions and suggests the inadequateness of the Heisenberg principle or an incompleteness of standard quantum theories. Thus, some unexpected *hidden variables* enter into play, revealing an uncommon structure of the universe that motivates harsh discussions among philosophers. My interpretation is based on the observation that these experiments are not conducted in a mathematical aseptic environment (in this utopian case, the results would be really embarrassing, nourishing the search for exotic explanations). On the contrary, particles interact with the electromagnetic background, especially near the measurement instruments. Thus, discrepancies on the outcomes are not to be ascribed to some exchange of information between particles, but to the effects of the entire environment that involves the use of material tools, with all the consequences discussed.

3.7 Implications on biological systems

Let me devote this short section to the crucial issue of correlating the results obtained here with some peculiarities of living systems. In particular, I would like to point out how the interpretation of a molecule as a gad-

---

[153] From [Boyer(1975)]: "Quantum theory, especially in its philosophical interpretations, is full of ideas about the influence of measuring apparatus upon the system being measured. This notion of unavoidable influence of the measuring apparatus is a natural deduction in our classical theory with zero-point radiation. The measuring apparatus will involve matter with electromagnetic interactions. This matter changes the pattern of zero-point radiation near the apparatus and so alters the system being observed".

[154] From [Little(1996)]: "In no way we are required to conclude that there is a breach between the real and the observed, between our knowledge and the objects of that knowledge. What we see is what exists. A measuring apparatus affects the creation of the particles that are observed; but this does not require the conclusion that a particle lacks identity until actually detected".

get, endowed with an inner dynamical individuality, might help understand biological processes[155]. I will limit the analysis to a few marginal considerations and ideas, due to my inexperience in these fields. In addition to the study of the constitution of living organisms, one may also wonder how significant is the impact of the presence of an organized electromagnetic background in our everyday life, and how this can affect our evolution, survival and connection with nature[156].

First of all I can quickly recall the effects (negative or positive) of exposure to daylight, that can affect biological molecules in direct or indirect ways. Molecules may be dissociated by photolysis, i.e., through the injection of energy carried by photons. For example, DNA may be severely damaged by UV light. Other stimuli are provoked by magnetic fields (see, e.g.: [Tenforde(1979)]) or electromagnetic noise ([Engels et al.(2014)]). In particular cases, magnetism can substitute gravitational effects on growing plants[157].

On the other hand, almost all living systems are equipped with chemical devices which transform visible photons into work[158]. In plants for example, the active core of chlorophyll consists of a porphyrin-like structure,

---

[155] From [Bajpai(2015)]: "The above implications lead us a possible answer of the age old question, who we are? The probable answer is: Each of us is a quantum entity entangled with a quantum photon field and both represent our individuality. The quantum entity is purely matter and localized while quantum field is purely energy and spread far and wide. The matter part is considered living and energy part, non-living but both contain equivalent amount of information about various aspects of 'life'. The two parts remain tuned with each other; any change in one part, is instantly reflected in other part as well".

[156] From [Moyroud et al.(2017)]: "Multiple wild species of bees have an innate preference for blue, which has been linked to the observation that flowers in the violet-blue range often produce relatively high volumes of nectar. However, blue color in petals is notoriously difficult to achieve. Flowers use sophisticated mechanisms to produce blue signals [...] we conclude that these floral nanostructures have converged on an optimized form that generates signals that are salient to insect pollinators".

[157] From [Geim(1998)]: "An interesting example of how the diamagnetic force can be exploited is an attempt to show that in space a magnetic field can replace gravity as a guide for plant growth: A germinating seed needs to know in which direction to grow so that it can successfully emerge from the soil before its limited resources are exhausted. Hasenstein's ground-based experiments indicate that even a small permanent magnet can provide enough guidance for a growing plant on board a spaceship".

[158] From [Horspool and Song(1955)], p.1412: "In plants, algae, and photosynthetic bacteria, the photosynthetic apparatus converts light energy into chemical energy. It has a complex structure that starts with an 'antenna' - an ensemble of pigments embedded in proteins, which absorbs light and funnel excitation energy toward specific proteins called reaction centers where photo-induced charge separation takes place".

made of Carbon and Nitrogen and presenting a sort of cavity, inside which a Magnesium atom is situated. The set resonates at various optical frequencies in the red and the blue spectrum. The energy of the absorbed photons is suitably driven and utilized to activate chemical reactions. The Magnesium is interpreted as Mg^{++} ion connected to two N^- ions. In my construction, these three atoms share a couple of electrons that occupy fixed positions between them. Such positions do not match those of the single isolated atoms but are individuated by the global shell system of all nuclei involved (see the case of nitric oxide discussed in section 3.3). The particular locations of these electrons force them to become sensitive to specific wavelengths. In the standard version, some free electrons then migrate through a series of molecules following specific cycles and participating in chemical reactions (see, e.g.: [Mathews and van Holde(1990)], chapter 19). Nevertheless, we know that an effective movement is not necessary, since it could be replaced by the right allowance of photons.

However, the aspect I am more concerned with is connected to the release of photons, since, following a generic multi-frequency injection of light, the photonic response of a molecule provides accurate information about its constitution. *Biophotonics* is a field of research that combines biology and photonics, which is the science that studies photons from the viewpoint of their generation, detection and manipulation (see for instance the early results in [Popp(2003)] and the review paper [Cifra, Fields and Farhadi(2011)]). With applications in life science, the discipline basically refers to photon emission and absorption by cells, but it is sometimes extended to tissues and even organisms. Biophotons usually range in the visible and ultraviolet spectrum. The detection of these photons is achieved through very sensitive equipment, and it is not to be confused with bioluminescence, where energy is released by living organisms through chemical reaction and successively in the form of light emission (as for example the *luciferin* emitted by fireflies). Most of the applications are in the medical field. The research is scattered in many independent branches, most of them sharing the study of the reactions of cells under laser stimulation. At the origin of the discipline is the discovery that all living systems permanently emit light quanta that induce chemical reactions in the cells. The cells themselves are a sort of cavity resonator contributing to biophoton regulation. The correct functioning of a living organism relies on the proper circulation of biophotons stored in the cells (see, e.g.: [Fleming(2014)]), so that one could in principle discriminate between healthy or cancer cells by differences in biophoton emission.

An intricate dynamical exchange of light quanta constantly occurs inside an organism, having the *DNA* molecules as principal points of absorption and emission. The activity of this communication network regulates, using proteins as intermediate transmitting stations, the most fundamental biological processes starting from morphogenesis, cell growth and cellular differentiation. Theoretically, information could be communicated not only throughout the organism but to the environment[159]. An affine branch of research is *optogenetics*, which combines optics with genetic manipulation. Here genetic engineering controls cell activity by illuminating photoreceptor proteins to change their conformation, thereby activating or inhibiting specific functions. The range of applications is extremely variegated and the results are promising.

Obviously, I am not surprised to learn about this mixture of complex chemical structures and photons, since the framework perfectly agrees with the one I developed starting from the simplest components. A great deal of biological processes are not simply due to straightforward chemical reactions. The truth is that messages carried by photons flow at the speed of light in a very organized electromagnetic landscape extending between atoms, supplying with lifeblood an inanimate scaffold of nuclei and electrons. Like in an electric circuit, both the various components and the currents supplied by a generator are essential parts of the machinery and, in a functioning device, they cannot be separated.

Hence, a DNA molecule is mainly an active apparatus governing the transfer of complex instructions in the form of photons, rather than just the sum of its nucleotides (see my version of cytosine in figure 3.19). Macromolecules ("cargo proteins") can cross back and forth the nucleus membrane in order to ensure the correct functioning of the cell. I believe however that information travels preferably via "radio waves", using massive molecules as intermediate transmitting stations, pre-constituted to resonate at specific

[159]From [Cifra, Fields and Farhadi(2011)]: "There is no doubt that biosystems can be affected by EMFs (ElectroMagnetic Fields) at several levels. There is also little doubt that biosystems can be the source of EMFs. The main question at hand is whether biosystems use EMF for a purposeful interaction (communication) and if so at what level of the bio-organism will it happen? The amount of data that support the latter notion is rapidly mounting at the same speed as the increasing number of questions that need to be addressed. [...] The prime question is 'Why we should care if cells interact via EMF?'. If the existence of distant cell communication proves to be true, there would be a substantial impact on our understanding of biology and biological research. Mastering and influencing the distant signaling system in biosystems can open a whole new horizon in our approach to biology. Then, the applications in biology and medicine could be astonishing".

signals, so avoiding the necessity of disturbing slow and inefficient chemical carriers.

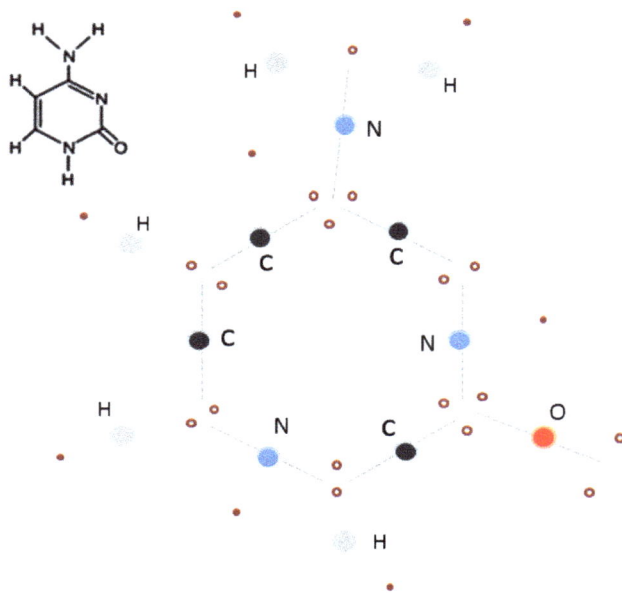

Fig. 3.19 *Tentative electron distribution of the cytosine molecule. Dashed segments denote the imaginary planes of symmetry of heavier atoms (C, N, O), as also shown in figures 3.10, 3.11 and 3.12. Single electrons are represented by small dots, whereas small annuli indicate that two electrons are lined up on an axis orthogonal to the page. Though the whole configuration is three dimensional, the molecule is basically planar. Close to the nucleus of each heavy atom there are two electrons (not shown), raising the total number of electrons to 58. Note that the nuclei are not at the vertices of the hexagon as standard chemistry notation suggests (see top-left scheme). The density of electron charge distribution of some nucleic bases has been computed in [Pullman, Dreyfus and Mély(1970)] by classical procedures. After removing the atom nuclei from those plots, similarities can be established with the picture above, by grouping the neighboring electrons. This somehow shows that electron displacement is very characterizing and my results are not in contrast with the existing ones, though obtained with quite different arguments. The atom structure is only a scaffold. The actual molecule is kept together by a vitalizing recirculation of photons. Some of them remain trapped inside, providing characteristic frequency patterns; some others are just passing by, influencing further molecules with their distinctive message.*

As the formation of a lifeless snowflake requires the simultaneous contribution of the water molecules to exchange information via the electromagnetic background (see section 3.4), more complex winding molecular systems may be formed on the same principles, that go beyond the minimization of the electrostatic energy[160]. Is this for instance the process at the base of the mechanism of *protein folding*?[161]

Furthermore, involved molecular aggregations may take advantage of the communication network to self-assemble and reproduce. If this is a viable explanation, certain proteins could undergo chemical synthesis from available C, N, O, H, atoms following recipes imprinted in the photon shells of the cellular environment. Assuming that nature can effectively account for my extended definition of chemical structure, which includes an impressive amount of well-organized interstitial photonic energy, I would be surprised if this feature was not wisely utilized to speed up evolution processes[162].

For instance, the question is to examine the role of existing elementary molecules (not very complex in terms of the number of constituting atoms, but already advanced in the evolution scale) displaying peculiar electromagnetic surrounding of the type examined so far and that are able to sort and redirect given basic atoms, helping their assembling with the aim of building other functional entities. Every time elementary building blocks of atoms join the system, the environment automatically changes to allow the selective access of an additional atom. The procedure stops, arriving at saturation, when a new molecule is created. Separation of the parent molecule from the generated one, would allow for the restarting of the guided cycle

---

[160] From [Leach(2001)], p.520: "One option is to parametrise the simple model to reproduce the results of a more detailed, all-atom model. An early attempt to develop such a representation was made by Levitt who used energy minimization to predict the structures of small proteins. [...] Some of Levitt's observations are still very pertinent. In particular, he noted that the 'wrong' structure may still have a lower energy than the 'correct' structure; this is also found to be the case with more complex molecular mechanics functions".

[161] From [Mathews and van Holde(1990)], p.197: "Although we still know only a little about the kinetics of protein folding, we are certain of one thing: folding does not proceed by anything approaching a random search through all the conformations possible to the unfolded form".

[162] From [Cosic et al.(2006)]: "The results from our acupuncture meridians and EEG activity studies confirm that the human body absorbs, detects and responds to ELF environmental EMF signals. This is a classical physics phenomenon utilized in telecommunication systems, which definitely needs to be further investigated for possible biological cell-to-cell communication phenomena".

(if other basic atoms are present to feed a new activity). *Catalysts* do actually work in the above mentioned fashion. The difference, according to my view, is that there is no need to move the reactant molecules physically, since the catalyst should be able to act from a distance, within its action range. This would be the *modus operandi* of the most basic reproduction machines — prototypes for much more evolved structures. The procedure should justify the preparation of cellular material of increasing complexity, perhaps in view of a cell duplication process. The trick is aimed at optimizing, in terms of processing time, those reproduction stages apparently ruled by randomness. In this way, the formation process is not only guided by crude deterministic chemical reactions, but follows an imperceptible scheme, prearranged by the organism's evolution history.

Instructive video simulations of excellent quality, displaying the interactions between DNA and proteins, tend to simplify the situation by showing the components flowing in an exceedingly spacious environment. In reality, in a cell, the space available is much reduced and completely filled with essential molecules. This significantly restricts the mobility of any massive components. The presence of other simple molecules, such as water, produces a screening effect, so that electrostatic forces are mitigated or redirected. On the contrary, photons are fast, they easily zigzag between atoms conveying precious information. Chain reactions can then occur with the active participation of proteins not primarily for their specific chemical constitution, but for the capacity of their atoms to influence the photonic environment. In this context, water is not an electrostatic impediment, but an indispensable mediator.

Let me stress again the concept of passing from a local behavior (the capability of a single cell to directly interact with the environment) to a global one (the capability of a union of cells to gather information from the outside and react together accordingly, even if the accomplishment seems to be pursued independently). For example, how can an entire part of a plant, not provided with a central nervous system, catch sensations at a very local level and respond with a collective behavior? Carnivorous plants are probably the most striking examples displaying such kind of reactions. Plants seem also to be able to communicate with each other through belowground interactions, while the results of events happen aboveground ([Elhakeem et al.(2018)]). A first quick answer is that nature is following the process of minimizing a certain global "functional" (using the language of mathematicians; see also at the end of section 3.4). However, even in the case of a single cell[163], the difficulty here is how to conceive the passage from a point-wise evaluation to a comprehensive output; a problem that in the end amounts to finding the explicit appropriate expression of the functional. My way to approach the problem would rely on the observation that an extended group of molecules may collectively "feel" the presence of an extraneous object because, even if the direct contact looks circumscribed, there are deeper interactions through the electromagnetic aura the various parts are embedded in. Similar arguments may be applied to many other circumstances[164]. Ants use trail pheromones as signaling technique to express social communications. How does an insect feel the presence of pheromones? By directly touching them? Improbable. By smelling the characteristic odor (as we do when sniffing the coffee aroma), even at large distance? Maybe, and in this case we already know how the machinery is activated. Pheromones are transmitting stations, electrically supplied by external agents. Infinitesimal lighted buoys for tiny navigators.

Moreover, we should not forget that the electromagnetic background extends above everything, with an immense range of frequencies that go far below the scale of biological phenomena, down to the slow motion of planets and further, to the imperceptible cycles of the evolving galaxies[165]. How might this affect life on Earth? I have no answer, but small traces of possible conflicts are scattered around, although discussing specific examples can be improper, due to the lack of unambiguous verifications. Let me just observe that unpredictable mutations in the mechanisms of genetic selection might be partly due to the undetectable influence of periodic phenomena

---

[163] From [Horspool and Song(1955)], p.1445: "Many freely mobile microorganisms like bacteria, unicellular algae, and protozoa are able to detect temporal variation in the external light field and to react to these environmental stimuli by modifying their movement, usually to achieve the best illumination conditions for their growth and metabolism and/or to avoid harmfully high light intensities".

[164] From [Bak, Tang and Wiesenfeld(1988)]: "On the other hand, it is well known that some dynamical systems act in a more concerted way, where the individual degrees of freedom keep each other in a more or less stable balance, which cannot be described as a perturbation of some decoupled state, nor in terms of a few collective degrees of freedom. For instance, ecological systems are organized such that the different species support each other in a way which cannot be understood by studying the individual constituents in isolation".

[165] From [Cravens(1997)]: "The *heliosphere* is a kind of bubble delimiting our Solar system. It is supposed to be composed by ionized light gasses, somehow "mixed" to magnetic fields. Astrophysical objects may interact with the surrounding flowing plasma to produce typical fluid dynamics structures, such as the so called *bow shocks*. This in general occurs near the boundary between a planet magnetosphere and the magnetized medium".

belonging to the large length scale. Evolution would then be the product of several concurring factors, intervening at different scales and reaching a complete symbiosis under the right circumstances[166].

There is evidence of cyclical behavior occurring in the universe, from the extreme low frequencies of galactic space to the periodic oscillations of a solar system. Although there is no direct proof of the influence of such slow cycles in human evolution, nobody can deny that this could be a possibility[167]. As I will better explain in the last section, coexisting overimposed recursive behaviors are all around us, ranging from *Brownian motion* of simple entities, passing to the frenetic agitation of insects, up to tidal waves. Each level may not be aware of being part of a lower frequency organization. Ants move fast not primarily for the reason that they are small, but because they had the chance to evolve in an environment ruled by certain high frequencies (compared to ours). Due to arguments of momentum conservation, physics relates high frequencies with small objects, and this lately explains why ants, if we assume them to be fast, must also be small.

Moreover, an insect's perception of the surrounding world is different from ours. Even if my hand moves fast, it is hard to catch a fly. A spider and a walrus are both supposed to belong to Earth's habitat; their perceived time flows however in different manners; to prove this, a rule of thumb is noticing that one of the two species is bigger than the other. The word "relativity" comes naturally into mind, but we can now give to it a deeper scientific meaning. The global universe is not a unique metric space but the union of a gigantic number of distinct and interlaced geometrical habitats, each one carrying a proper clock (see my comments at the end of chapter 1). Einsteinian relativity is now at its maximum stage of potentiality, and here the theory meets *relativism*.

---

[166] From [Gould(2002)], p.32: "Thus, for Darwin's near exclusivity of organismic selection, we now propose a hierarchical theory with selection acting simultaneously on a rising set of levels, each characterized by distinctive, but equally well-defined, Darwinian individuals within a genealogical hierarchy of gene, cell-lineage, organism, deme, species, and clade. The results of evolution then emerge from complex, but eminently knowable, interactions among these potent levels, and do not simply flow out and up from a unique casual locus of organism selection".

[167] From [Dewey and Dakin(1947)]: "The student of periodic rhythms in human affairs has a tool which the law of averages itself puts into his hands. If trends have continued for decades, or if the oscillations of cycles around the trend have repeated themselves so many times and so regularly that the rhythm cannot reasonably be the result of chance, it is unwise to ignore the probability that these behaviors will continue".

3.8   The birth of an idea

Physical phenomena at the nuclear and atomic level are dominated by very high frequencies. Conversely, astronomical and cosmological events are ruled by very low frequencies. In the middle, depending on the magnitude of the objects under study, we find a whole range of frequencies, roughly connecting the objects' size with the inverse of the frequency[168]. There is a sort of scalability in these processes, so that by enlarging the scene of some event, this becomes more credible if we also slow down its velocity[169]. For example, a trick often used by movie makers to obtain impressive special effects from miniaturized models is the following: in order to make things look real, the movie is shot at high speed and then played in slow motion. The nanomovies of A.H. Zewail (see, e.g., [Zewail(2010)]) are wonderful upshots of this modus operandi. Implicitly, this has also a scaling effect on the perception of masses, and shows that, based on our experience, we can more or less guess the magnitude of a mass from the speed and the way it moves, without even touching it. We know that an eagle will never be able to flap its wings as fast as a bumble-bee. Nevertheless, a magnified and decelerated bee will appear as a plausible threatening monster. The fact that in a movie we recognize large slow massive objects to be real, when instead they are just the magnification of small fast light models, tells us how precarious is the concept of mass. Such a promiscuity between size and frequency seems to exist before the idea of mass is actually introduced. This observation follows an inverse path of that pursued for instance in classical mechanics, when the period of a pendulum is calculated starting from the weight force. Oscillations come first; then other concepts may follow. Concurrently, things start assuming a fractal-like structure, where big objects are linked to their smaller constituents by scaling transformations that also reparametrize time.

Maybe, one may find relationship with *string theories*, where for instance electrons are mono-dimensional vibrating segments. There, the ba-

---

[168]From [Kaku(1991)], p.4: "The myriad particles found in nature can be viewed as the vibrations of a string, in much the same way that the notes found in music can be explained as the modes of a vibrating string. Pursuing this analogy, the basic particles of our world correspond to the musical notes of the superstring, the laws of physics correspond to the harmonies that these notes obey, and the universe itself corresponds to a symphony of superstrings".

[169]From [Shikhmurzaev(2008)], p.9: "Indeed, on a geological time and length scale the mountains are quite 'fluids', whilst a microscopic droplet of water on a microscopic time scale can behave like a solid particle".

sic assumptions are oversimplified but they soon develop in a very complex abstract way, to the point that they may require many more space dimensions than the usual. String theorists are able to merge quantization with general relativity, an achievement that is also pursued in these notes. The final conclusions, though reached following a different path, are not dissimilar[170]. A strong attack on string theories has recently been brought to the attention of the public. The criticism recognizes the ideals of the theory and the elegant mathematics, but severely relegates it to be a mere theoretical effort with no practical content[171]. It is not my desire however to enter the debate. Let me just point out that string theories give the possibility of reinterpreting Feynman's diagrams, used in the description of subatomic interactions, in terms of topological changes at string level. There might not be any direct resemblance with my 3D evolutive model, where boiling bubbles, shells and layers, fight at every instant for living space and eventually bifurcate or disappear with a pop. However, if I was an expert I would search for some affinity. For example: is there any relationship between my uncharged neutrino and the closed massless *graviton*?

Animated by the above considerations, I started my review of the foundations of physics, and here is a little resumé of the developing phases. The real goal was to face the problem of giving a rational sense to quantum phenomena and to the constitution of basic atoms. The fact that there are frequencies hidden inside an atomic structure is inspiring. If oscillations come first, they approximately have to span a region of space of magnitude inversely proportional to the frequency. Are atomic and molecular bonds a consequence of resonant properties of such a pre-existing set of oscillators? That was roughly the idea, but to be acceptable, I had to specify what was

---

[170] From [Kaku(1991)], p.4: "Moreover, when the string executes its motions, it actually forces space-time to curl up around it, yielding the complete set of Einstein's equations of motion. Thus, the string naturally merges the two divergent pictures of a force: the modes of vibration are quantized, but the string can only self-consistently vibrate in a curved space-time consistent with Einstein's equations of motion".

[171] From the introduction in [Woit(2006)]: "The willingness of some physicists to give up on what most scientist consider the essence of the scientific method has led to a bitter controversy that has split the superstring theory community. Some superstring theorists continue to hold out hope that a better understanding of the theory will make the landscape problem go away. Others argue that physicists have no choice but to give up on long-held dreams of having a predictive theory, and continue to investigate the landscape, hoping to find something about it that can be used to test an idea experimentally. The one thing both camps have in common is a steadfast refusal to acknowledge the lesson that conventional science says one should draw in this kind of circumstance: if one's theory can't predict anything, it is just wrong and one should try something else".

actually oscillating and based on what kind of rules. I considered the first question secondary and I tried to concentrate my attention on a mathematical model able to fulfill the following request: the wave emanated by some fixed source has to dampen with distance, not only in amplitude, but also in frequency. Through such a mechanism, it turns out that the "sound" has to be high pitched in proximity to a given source, but the tonality has to lower in moving away. The phenomenon is indeed nonlinear.

Without bothering about applicability issues, I began to play with Maxwell's equations, just to have a solid starting point to develop a non-linear set of partial differential equations. The first edition of the modified model was not too difficult to obtain and seemed quite promising. Simultaneously, I understood that there was something wrong in the Maxwell's model. Although I tried to act prudently, the inconsistency remained. At that stage, my fight for a more coherent description of electromagnetic phenomena began, with all the consequences described in this book, including the most important one: the unifying connection between corpuscular and undulatory theories[172]. It is for me important to recall that I believed from the very beginning in a quantitative description based on old-fashioned differential tools. Therefore, the systematic algebraic approach of the *standard model* and the strict geometric attitude of *string theories* were not suitable.

In conclusion I got what I anticipated at the beginning of the present section: an oscillating background with an extended range of frequencies, and together with it, a preparation to understanding the meaning of mass. Everything is described by deterministic field equations, seemingly up to the last detail. The whole discovery process was not painless, since I had to overcome many clichés. I found myself reviewing and connecting too many different branches of knowledge, an enterprise hard to accomplish with the necessary depth by a single person[173]. I think however that the final result,

---

[172] In [Planck(1927)] the author claims: "In what relation, however, the corpuscular laws stand to the laws of wave-motion in the general case, remains the great problem, to which at present time a whole generation of investigators is devoting its best efforts. We can entertain no doubt that finally a satisfactory solution will be found, and that then theoretical physics will have made another significant advance toward the attainment of its ultimate goal, the building up of a unified system embracing all physical phenomena".

[173] From [Maruani(1988)], v.1, p.228 (G. Naray-Szabo): "The ambition to achieve interdisciplinarity may not mean that the rigorous treatment of a given problem should be abandoned. Clearly, careless work, improper argumentation or logical errors cannot be tolerated in a scientific study. Interdisciplinarity therefore involves profound studies in the related areas that are not always practicable for a single person. Team-work is necessary ...".

although in need of confirmation and some interpretational adjustment, is quite consistent and pragmatic. I wonder if my unconventional claims can alter the conservative attitude of the contemporary science[174].

Only lately, I realized that the frequency decay with distance cannot be achieved continuously but requires quantum jumps. Starting from nuclei, where the highest frequencies are concentrated, a sequence of nested shells having geometrical growth is found. If one shell is removed, perhaps because it has been perturbed from the exterior with the due resonance frequency, the neighboring shells tend to fill the gap. In doing this they adapt in size and frequency to restore the missing one.

In [Bak, Tang and Wiesenfeld(1988)], the authors consider the dynamical behavior of a sand-pile where additional sand is supplied from above. By analyzing what they define as *self-organized criticalities*, it is found out that the energy spectrum radiated from this system can be put in relation to *flicker noise*, where intensity is proportional to a given power β of the inverse of the frequency (we have *pink noise* when $\beta = 1$). It is also argued that many physical phenomena, connected with self-similar fractal structures, display such noise production, although the origin of this property is unknown[175]. Predictions are finally confirmed by numerical tests. We can now look at my shell agglomerates as kinds of inverted sand-piles, i.e., sand holes where sand disappears when it gets to the bottom, as in the upper part of an hour glass. Note that in this case the sand distribution at the surface does not have the shape of a reversed cone, being instead more similar to something amenable to an exponential. After better formalization of the problem, it would then be interesting to perform an analysis to check if noise production from this system is still of flicker type. Such a study might help to determine the reasons for the ubiquity of this kind of noise in nature's events and provide another confirmation of my approach.

Let me add further philosophical thoughts. Throughout the book, I

---

[174] From [Cantore(1969)], p.230: "Physicists tend to be conservative. Despite the impression to the contrary frequently conveyed by popular writings, this is a historical fact. The numerous controversies that constantly accompanied atomic physics during its formative years are evidence enough that physicists tend to cling to currently prevailing conceptions and greet innovations with skepticism or downright opposition. And this is quite reasonable. Earnest researchers cannot be expected to renounce established ideas once they have proved their validity. The trouble is that the same scientists may extrapolate their already demonstrated conclusions into unexplored domains of physical reality".

[175] From [Bak, Tang and Wiesenfeld(1988)]: "To summarize, our general arguments and numerical simulations show that dissipative dynamical systems with extended degrees of freedom can evolve towards a self organized critical state, with spatial and temporal power-law scaling behavior. The spatial scaling leads to self-similar fractal structure".

always used the term "model" when referring to the proposed set of equations. Nowadays, within the scientific context, the word model corresponds to a system of axioms, able to reproduce some specific processes up to a certain degree of approximation. In physics, the model should describe, through the language of mathematics and as realistically as possible, how natural phenomena work. There is however another way to interpret the concept: a model is a theoretical prototype which underlies reality, like the design drawn by a tailor that eventually becomes a dress. In this way a model does not explain a given reality, but anticipates it. The dilemma is the following: is the model here described an accurate description of a given "real" world? Or, is it a set of recipes, useful in the construction of a possible real world, as the one we live in? Answering the second question would be quite an ambitious project. The idea of coding natural laws with the help of a few basic mathematical formulas has been the dream of many scientists. Even more appealing was the hope that the goal was not just a description, but a series of primitive instructions from which everything originated.

It has to be remarked that if a theory is to be as simple as possible, it needs a huge quantity of elements to operate with. This seems not to be a problem since the universe is large enough. In addition, even the most elementary significant structure, such as the bare electron, requires a very involved description. All the equations are fully used to describe a bare particle and this somehow confirms the minimalism of the assumptions. Once the basic building blocks are available, the rest becomes an arduous technical exercise that, due to the simplicity of the governing laws, allows for an immense variety of solutions. Electromagnetism generates particles that, with the help of the same vitalizing source, may give rise to clusters of molecules, even leading to the origin of life. We would then be in an amazing situation in which the mathematical solution of a set of partial differential equations acquires self-consciousness of its own originating model. Whether or not my model is correct, this last observation is a shocking remark. The problem remains as the identification of what initial conditions can lead to this and how the constants involved are crucial.

I believe that the universe is a kind of fancy self-generated[176]

---

[176] From [Wheeler and Zurek(1983)], p.205 (J.A. Wheeler): "The absolute central point would seem to be this: The Universe had to have a way to come into being out of nothingness, with no prior laws. no Swiss watchworks, no nucleus of crystallization to help it — as on a more modest level, we believe, life came into being out of lifeless matter with no prior life to guide the process. When we say 'out of nothingness' we do not mean out of the vacuum of physics. The vacuum of physics is loaded with geometrical

"homework" and, accidentally, we are part of it, whether or not we observe and understand its rules. Before the effective realization, there is the "plan" (whatever this vague concept may mean) and the constituting laws. Based on the same identical laws, things could have been evolved in a different manner.

At larger scales, the explanation of fundamental facts in the framework of a theory that originates from electromagnetic fields is certainly too complex. In this case offhand phenomenological conclusions could turn out to be more effective[177]. Finding a direct link between atomic phenomena and celestial mechanics is a rather involved process necessitating several intermediate steps where macro-agglomerates of increasing complexity are fused and treated as single entities subjected to revised laws. New fundamental constants may emerge in this averaging process[178,179]. Nevertheless, as I said in the preface, the original goal was to provide an overall meaning to the greatest number of phenomena, although global optimization is not necessarily the best choice at a local level. Thus, simplicity might be sacrificed in particular cases in favor of a unifying vision[180].

---

structure and vacuum fluctuations and virtual pair of particles. The Universe is already in existence when we have such a vacuum. No, when we speak of nothingness we mean nothingness: neither structure, nor law, nor plans".

[177]From [Shikhmurzaev(2008)], p.10: "The reason why the fluids which are organized so differently behave qualitatively in the same way when observed on a macroscopic length and time scale is still not quite clear. It is even less clear why the chain of casual links between processes on different length and time scales, from the macroscopic down to the molecular level, can be broken to form a closed set of equations involving a small number of parameters on the macroscopic level with the microscopic 'tail' replaced by a few, in many cases constant, coefficients (e.g., viscosity, thermal conductivity, etc.)".

[178]From [Bohm and Hiley(1982)]: "For example, several quantum systems may separate in such a way that their wave functions factorize, so that they behave independently. Yet, later, such systems may combine again with each other and with other systems to form new quantum states of a single larger system. When this happens, the systems that were independent now cease to be so, and a new quantum potential arises, in which the interactions between particles within any one of the initially independent systems are now found to be dependent on the state of the whole larger system in which they take part".

[179]From [Gallavotti(2002)], p.442: "This immediately makes us understand that is should be possible to express the phenomenological coefficients of viscosity or thermal conductivity in terms of *averages*, over time and space, of microscopic quantities which are more or less rapidly fluctuating. We deduce that transport coefficients (such as viscosity or conductivity, etc.) *do not have a fundamental nature*: they must be rather thought of as macroscopic parameters that measure the disorder at molecular level".

[180]From [Elitzur, Dolev and Kolenda(2005)], p.3: "One of the most profound features of reality is that simplicity goes hand in hand with universality. One may drop a basic assumption or even an axiom and, lo and behold, the edifice built on the remaining

On a qualitative level, the analysis of the micro-world, completely filled up by zillions of evolving geometrical units carrying energy, may teach us something new about the macro-world. The work done here to unify electromagnetic and mechanical forces is at present not sufficient to understand the gravitational attraction of bodies. Throughout the exposition it has been remarked that the role of the magnetic field is in assigning mass to particles and justifying the constitution of atoms. Therefore, I would not be surprised if one discovers that magnetic forces are essential for explaining gravitational phenomena[181]. Moreover, let me also observe that the vision of the greater universe from our standpoint has a chance to be affected by aberrations, due to the filtering of the incoming signals in the passage through invisible surrounding shell boundaries. The redshift effect, mentioned at the end of section 1.4, might be the consequence of distorted information that, akin to a huge kaleidoscope, induces us to see patterns in the space-time not corresponding to what is supposed to be reality.

At this point physicists may ask for some predictive hypotheses aimed at validating the theory. Though there are some ideas that I would like to explore, I do not really believe that it is compulsory to look for new experimental results; in the end my goal was to point out the deficiencies of existing theories and find out the way to link various aspects of them based on known empirical facts. The complicated world of quantum phenomena, far from classical principles as it is, and the surprising variety of nuclear phenomena instill in us the idea that an epistemological explanation may not actually exist or has to be based on tools yet to be invented. Now we are standing in front of a new proposal whose actors are known (the vector fields). The rules and the properties are those developed through centuries of hard science[182], and the results are quite in accordance with observation, without needing further postulates. Is this the instruction set of our universe? It would be enough to show at least an unexplainable fact

---

narrower foundations turns out to be *wider*: additional phenomena, beyond those which one sought to explain, turn out to fit neatly within the new theory".

[181] From the Preface in [Clauser(1960)]: "More and more astrophysicists are recognizing that magnetic fields are a powerful agent in the dynamics of our universe. Our own earth, sun, planetary system, stars, and the material between them are all profoundly influenced by magnetic fields".

[182] From the introduction in [Newton(2009)]: "If the purpose of physics is to correctly describe nature, then two of the various aspects of this description are among the most important: what the constituents of the world are and how objects move. We want to know what the world consists of and what accounts for the changes we constantly observe, that is, what are the dynamical laws underlying these motions".

or a non-matching constant to destroy the whole dream. In chapters two and three, my exposition was mainly qualitative because of the difficulty of solving the model equations in such complicated situations. Therefore, the possibility of misconception and overestimation is high. The reader cannot deny however that the achievements presented here are intriguing and provide us with an alternative way of deciphering our universe[183].

---

[183] From the *The Hitchhiker's Guide to the Galaxy* by Douglas Adams: 'Forty-two,' said Deep Thought, with infinite majesty and calm. [...] 'I checked it very thoroughly,' said the computer, 'and that quite definitely is the answer. I think the problem, to be quite honest with you, is that you've never actually known what the question is'.

Chapter 4

Appendices

Esistono delle leggi naturali di una profondità
e di una bellezza incredibili
Non si può pensare che tutto ciò
si riduca ad un accumulo di molecole
Carlo Rubbia, scientist

A   Maxwell's equations in vacuum

The main properties of the Maxwell's system of equations in vacuum are going to be collected here. Denoting the electric field by $\mathbf{E} = (E_1, E_2, E_3)$ and the magnetic field by $\mathbf{B} = (B_1, B_2, B_3)$, we first have the *Ampère's law*, with no current source term:

$$\frac{\partial \mathbf{E}}{\partial t} = c^2 \text{curl} \mathbf{B} \tag{1}$$

where $c \approx 2.99 \times 10^8$ m/sec denotes the speed of light. Successively, we have the *Faraday's law of induction*:

$$\frac{\partial \mathbf{B}}{\partial t} = -\text{curl} \mathbf{E} \tag{2}$$

and the two following conditions on the divergence of the fields:

$$\text{div} \mathbf{E} = 0 \tag{3}$$

$$\text{div} \mathbf{B} = 0. \tag{4}$$

Multiplying (1) by $\mathbf{E}$ and (2) by $\mathbf{B}$, summing up and using basic notion of differential calculus, one gets:

$$\frac{1}{2} \frac{\partial}{\partial t} (|\mathbf{E}|^2 + c^2 |\mathbf{B}|^2) = -c^2 \text{ div}(\mathbf{E} \times \mathbf{B}). \tag{5}$$

Note that the quantity $\mathcal{E} = \frac{1}{2}(|\mathbf{E}|^2 + c^2|\mathbf{B}|^2)$ is proportional to the density of electromagnetic energy.

By suitably combining the equations (1), (2), (3), (4), it is not difficult to arrive at the wave equations:

$$\frac{\partial^2 \mathbf{E}}{\partial t^2} = c^2\,\Delta\mathbf{E} \qquad \frac{\partial^2 \mathbf{B}}{\partial t^2} = c^2\,\Delta\mathbf{B}. \qquad (6)$$

It is standard to introduce the electromagnetic potentials Φ and $\mathbf{A} = (A_1, A_2, A_3)$, such that:

$$\mathbf{B} = \frac{1}{c}\,\mathrm{curl}\mathbf{A} \qquad \mathbf{E} = -\frac{1}{c}\frac{\partial \mathbf{A}}{\partial t} - \nabla\Phi. \qquad (7)$$

By assuming this, equations (2) and (4) are automatically satisfied. The potentials are not unique, but are usually related through some gauge condition. For convenience, the following *Lorenz gauge* will be assumed:

$$\mathrm{div}\mathbf{A} + \frac{1}{c}\frac{\partial \Phi}{\partial t} = 0. \qquad (8)$$

The writing of Maxwell's equations in tensor form is obtained as follows. Using the notations in [Jackson(1975)], in the system of time-space Cartesian coordinates $(x_0, x_1, x_2, x_3) = (ct, x, y, z)$, one first defines the electromagnetic tensor:

$$F_{\alpha\beta} = \partial_\alpha A_\beta - \partial_\beta A_\alpha \qquad (9)$$

where ∂_α is the α-component of the differential operator $(\partial/\partial x_0, \nabla)$ with $\nabla = (\partial/\partial x, \partial/\partial y, \partial/\partial z)$ and A_α is the generic entry of the tensor $(\Phi, -\mathbf{A})$. According to (7), the anti-symmetric electromagnetic tensor and its *contravariant* version take the explicit form:

$$F_{\alpha\beta} = \frac{\partial A_\beta}{\partial x_\alpha} - \frac{\partial A_\alpha}{\partial x_\beta} = \begin{pmatrix} 0 & E_1 & E_2 & E_3 \\ -E_1 & 0 & -cB_3 & cB_2 \\ -E_2 & cB_3 & 0 & -cB_1 \\ -E_3 & -cB_2 & cB_1 & 0 \end{pmatrix} \qquad (10)$$

$$F^{\alpha\beta} = g^{\alpha\gamma}g^{\beta\delta}F_{\gamma\delta} = \begin{pmatrix} 0 & -E_1 & -E_2 & -E_3 \\ E_1 & 0 & -cB_3 & cB_2 \\ E_2 & cB_3 & 0 & -cB_1 \\ E_3 & -cB_2 & cB_1 & 0 \end{pmatrix} \qquad (11)$$

with $g^{\alpha\beta} = g_{\alpha\beta} = \mathrm{diag}\{1, -1, -1, -1\}$. According to Einstein's notation, in (11) one has to sum up from 0 to 3 on repeated indices.

Maxwell's equations (1) and (3) are then obtained by computing:

$$\frac{\partial F^{\alpha\beta}}{\partial x_\alpha} = 0 \qquad \text{for } \beta = 0, 1, 2, 3 \tag{12}$$

where one has to sum up on the index α. As a matter of fact, for $\beta = 0$ one gets (3), and for $\beta = 1, 2, 3$ one gets the three components of (1). The above expression can be shortened as follows:

$$\partial_\alpha F^{\alpha\beta} = 0. \tag{13}$$

The electromagnetic *Lagrangian density* assumes the form:

$$\mathcal{L} = -F_{\alpha\beta}F^{\alpha\beta} = 2(|\mathbf{E}|^2 - c^2|\mathbf{B}|^2). \tag{14}$$

The search for the stationary points of the *action function* associated to the Lagrangian, by applying all the possible variations δA_α to the potentials, leads to equation (13). In fact, after carrying out computations that for simplicity are not reported here, one finds that:

$$(\partial_\alpha F^{\alpha\beta}) \, \delta A_\beta = 0 \qquad \forall \, \delta A_\beta. \tag{15}$$

B   On the divergence of the electric field

An accelerated charged body produces electromagnetic radiation. There are circumstances in which this process is not very clear from the viewpoint of the classical equations of electromagnetism, as shown here through simple arguments. First of all, we need to assume that the information propagates at finite speed, i.e., that of light. At rest, the body is surrounded by its own stationary electric field $\mathbf{E}$. If the charge is concentrated, we automatically have $\rho = \mathrm{div}\mathbf{E} = 0$ everywhere, with the exception of the point carrying the charge. By moving the particle from its initial position, the field changes, but, at the beginning, this only occurs in a near region that successively grows, propagating to all the space around[184]. Thus, for a small fraction of time, there are far-away zones where the movement of the particle has not been felt yet. An attempt to describe the dynamics of $\mathbf{E}$ for an accelerated charge is provided in figure 4.1. The picture is very similar to that shown in [Eisberg and Resnick(1985)], appendix B.

---

[184] From [Landau and Lifshitz(1961)], p.47: "A change in the position of one of the particles influences other particles only after the lapse of a certain time interval. This means that the field itself acquires physical reality. We cannot speak of a direct interaction of particles located at a distance from one another. Interactions can occur at any one moment only between neighboring points in space (contact interaction). Therefore we must speak of the interaction of the one particle with the field, and of the subsequent interaction of the field with the second particle".

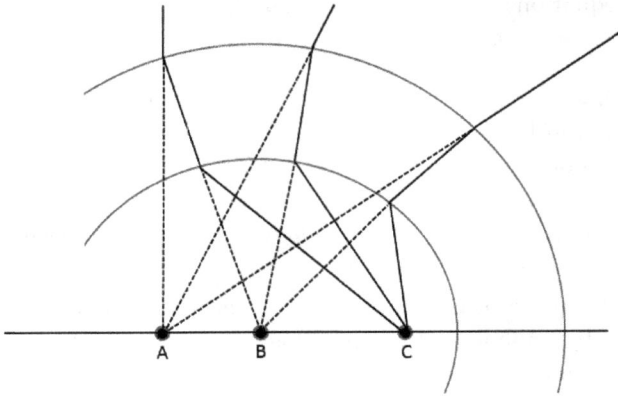

Fig. 4.1 *Lines of force of the electric field, when a single charge is accelerated from a position at rest A. In the far-away region the distribution coincides with that corresponding to the charge still located in A. In the intermediate region, the lines are adjusted as the charge is concentrated in B. In reality, this process is continuous, and this visualization at different steps only serves to make the exposition clear The paths of the lines of force are heavily distorted, since we are supposing that in C the particle travels at almost the speed of light. Of course, the curvature is far less pronounced when the particle moves at lower velocities.*

The case of a single point-wise moving particle can be handled with the help of the standard Maxwell's model, by adding the corresponding current term (of singular type) on the right-hand side of equation (1). There are however special situations that are not covered by the theory. Here are some examples. Suppose that the charge, uniformly distributed on a sphere of radius $\delta > 0$, is pulsating. This produces a radial electric field that in spherical coordinates takes the expression: $\mathbf{E} = (r^{-2} \sin \omega(ct - r), 0, 0)$, for $r \geq \delta$. Everywhere, we have: curl$\mathbf{E} = 0$ and $\mathbf{B} = 0$ (indeed, for symmetry reasons there is no generation of magnetic field). Therefore, (2) is satisfied. Nevertheless, we have $\rho = -r^{-2} \cos \omega(ct - r) \neq 0$, causing the generation of spherical density waves that escape from the source at speed c. Now, (1) is not correct anymore, but it can be replaced by the equation $\partial \mathbf{E}/\partial t = -\rho \mathbf{V}$, with $\mathbf{V} = (c, 0, 0)$, which is the same introduced at the very end of appendix G. Note that, at the points where $\rho \neq 0$, there are no real physical charges, according to the definition generally adopted. Another case is when the sphere maintains its global charge but changes its diameter. If the surface of the sphere is a perfect conductor, the intensity of the electric field instantly adapts itself to the new curvature, so that the external field remains compatible with the stationary solution $\mathbf{E} = (r^{-2}, 0, 0)$, having $\rho = 0$. Since perfect conductivity is just a mathematical abstraction, a real situation requires a delay in the rearranging procedure, with corresponding production of regions where $\rho \neq 0$. One can then imagine more involved contexts, where charged bodies move and change shape at the same time. We are allowed to conclude that the "inertia" of the surrounding vacuum in reaction to sudden modifications, may give rise to incompatibilities with the equation $\rho = 0$.

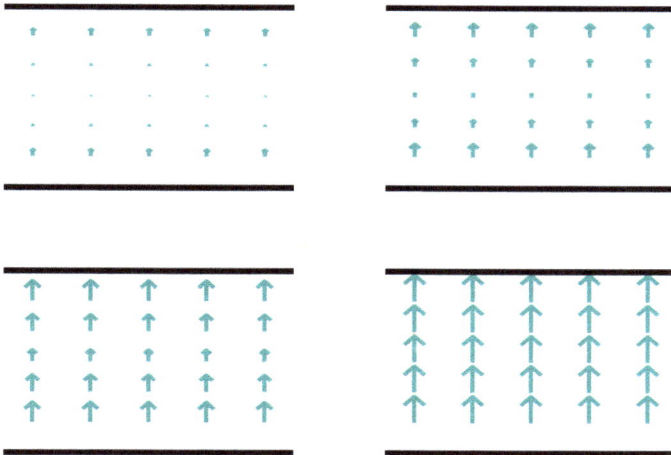

Fig. 4.2 *Displacement of the electric field* $\mathbf{E}$ *during the charge of a capacitor (large plates at small distance). As the difference of potential between the plates increases, the divergence ρ remains different from zero, until the process is concluded. The discrepancies smooth out quite fast, so that the effect can be neglected in many applications. This does not mean that, in general, the problem can be underestimated.*

There are strong mathematical arguments in support of the fact that the wave equation for $\mathbf{E}$ and the divergence free condition $\rho = 0$ cannot hold at the same time, unless we are at very specific regimes. In [Engelhardt(2012)], the author points out contradictions in solving Maxwell's equations in the case of a single moving charge. Indeed, different conclusions may be achieved depending on the way the problem is approached. In [Zhakatayev(2016)], theoretical computations based on the retarded potential show that for accelerated non-singular charges at relativistic velocities, we can actualy expect regions where the divergence of the electric field is not zero. In [Funaro(2008)] it is argued that waves with spherical fronts (or other types of surfaces developing with the rules of geometrical optics) must display zones having $\rho \neq 0$. In truth, optical waves are compatible with Maxwell s equations only by assuming suitable approximations (see, e.g.: [Born and Wolf(1980)]). Again, due to the finiteness of the speed of light, when high-frequency electric signals are applied to a capacitor, the information between the walls propagates with an appreciable delay (see figure 4.2). This is typical of capacitors having the distance between the plates larger than the wavelength of the signal, which is the situation one may encounter, for instance, in the near-field of antennas. Therefore, even in vacuum, the fast variation of the signal produces sinks and sources in the electric field, giving rise to regions where the divergence is not zero. The classical Coulomb's law and Maxwell's equations do not correctly describe this behavior. As far as the case of a simple capacitor is concerned, the theoretical analysis carried out in [Funaro(2014)] rigorously shows that enforcing the full set of Maxwell's equations actually leads to incompatibilities. In particular, the wave equation for the electric field enters in conflict with the divergence-free condition for a wide range of boundary conditions. By adopting the revised model given in appendix G, a more consistent description of the development of the fields inside a capacitor is achieved ([Funaro(2018b)]).

C   The model equations for free-waves

We introduce the equations studied in [Funaro(2005)] and [Funaro(2008)]. In the simple case of *free-waves*, they are as follows:

$$\frac{\partial \mathbf{E}}{\partial t} = c^2 \text{curl}\mathbf{B} - \rho\mathbf{V} \tag{16}$$

$$\frac{\partial \mathbf{B}}{\partial t} = -\text{curl}\mathbf{E} \tag{17}$$

$$\text{div}\mathbf{B} = 0 \tag{18}$$

$$\rho\left(\mathbf{E} + \mathbf{V} \times \mathbf{B}\right) = 0 \tag{19}$$

where by definition: $\rho = \mathrm{div}\mathbf{E}$. Here $\mathbf{V} = (V_1, V_2, V_3)$ is a velocity field, built in such a way that the triplet $(\mathbf{E}, \mathbf{B}, \mathbf{V})$ is right-handed. Equation (16) turns out to be the Ampère law for a free flowing immaterial current having density ρ associated with the movement of the electromagnetic wave. From (19), one easily gets the orthogonality relations: $\mathbf{E} \cdot \mathbf{B} = 0$ and $\mathbf{E} \cdot \mathbf{V} = 0$. This allows us to prove again relation (5), after multiplication of (16) by $\mathbf{E}$ and (17) by $\mathbf{B}$. In addition, we assume that the intensity of $\mathbf{V}$ is constantly equal to the speed of light, i.e.:

$$|\mathbf{V}| = c. \tag{20}$$

In particular, if $\mathbf{V}$ turns out to be the gradient of a potential function Ψ, equation (20) becomes:

$$|\nabla\Psi| = c \tag{21}$$

which is the stationary *eikonal equation*.

A way to satisfy the geometric relations (19) and (20) is to set $\mathbf{V} = c(\mathbf{E} \times \mathbf{B})/|\mathbf{E} \times \mathbf{B}|$. Thus, from (19) one also gets $|\mathbf{E}| = |c\mathbf{B}|$. These relations actually hold in many circumstances (see some exact solutions in appendix E). In these situations the triplet $(\mathbf{E}, \mathbf{B}, \mathbf{V})$ is orthogonal.

Since a current term is present, due to preservation arguments, a *continuity equation* must hold. This does not need to be imposed independently, since it is easily obtained by taking the divergence of (16):

$$\frac{\partial\rho}{\partial t} = -\mathrm{div}(\rho\mathbf{V}). \tag{22}$$

One can still define the potentials as in (7) and the electromagnetic tensors as in (10) and (11). With this setting, equations (17) and (18) are automatically satisfied.

In tensor form, equation (16) is now written as:

$$\frac{\partial F^{\alpha\beta}}{\partial x_\alpha} = \frac{1}{c}\frac{\partial F^{\gamma 0}}{\partial x_\gamma} V^\beta \qquad \text{for } \beta = 1, 2, 3 \tag{23}$$

where $V_\alpha = (c, -\mathbf{V})$ and $V^\beta = g^{\beta\delta}V_\delta = (c, \mathbf{V})$. It is worthwhile recalling that one has to sum up on repeated indices. Note that, for $\beta = 0$, (23) turns out to be a trivial identity, hence equation (3) is not actually imposed. We also have:

$$V^\alpha V_\alpha = 0 \tag{24}$$

that corresponds to (20).

Finally, the whole set of equations in compact form takes the form:

$$\partial_\alpha F^{\alpha\beta} = \frac{\rho}{c} V^\beta \tag{25}$$

$$F^{\alpha\beta} V_\beta = 0 \tag{26}$$

with $V^\alpha = (V^0, \mathbf{V})$. Indeed, up to a multiplicative scaling factor, it is not restrictive to choose: $V^0 = V_0 = c$, as suggested by condition (20). Therefore, by examining (25), for $\beta = 0$ one gets $\rho = \mathrm{div}\mathbf{E}$, while for $\beta = 1, 2, 3$ one gets the three components of (16). Concerning the tensor multiplication in (26), for $\beta = 0$ one finds that $\mathbf{E}$ must be orthogonal to $\mathbf{V}$, while for $\beta = 1, 2, 3$ one gets the three components of (19).

A Lagrangian for the new formulation is obtained by taking $\mathcal{L} = 2(|\mathbf{E}|^2 - c^2|\mathbf{B}|^2)$, i.e., the same as the classical Maxwell's case (see (14)). In order to get equation (16), it is necessary however to fix a constraint on the potentials. This is given by:

$$c\mathbf{A} = \Phi\mathbf{V} \tag{27}$$

which says that $\mathbf{A}$ must be lined up with $\mathbf{V}$. For an explanation of this choice one may look in [Funaro(2008)], section 2.4, and in [Funaro(2010)]. The exact solutions given in appendix E actually satisfy such a restriction.

From (27) and (20), one gets $|\mathbf{A}|^2 = \Phi^2$ and $\mathbf{A} \cdot \mathbf{V} = c\Phi$, that in tensor form read as follows:

$$A_\alpha A^\alpha = 0 \quad \text{and} \quad A_\alpha V^\alpha = 0 \tag{28}$$

having set $A_\alpha = (\Phi, -\mathbf{A})$. By applying such a restriction to the variations δA_α one finds out that $V^\beta \delta A_\beta = 0$. Therefore, relation (15) must be replaced by:

$$(\partial_\alpha F^{\alpha\beta}) \, \delta A_\beta = 0 \quad \forall \, \delta A_\beta \ \text{such that} \ V^\beta \delta A_\beta = 0. \tag{29}$$

This says that $\partial_\alpha F^{\alpha\beta}$ and V^β are linearly dependent, which leads to equation (25).

Finally, by taking the time-derivative of (16) and using (22), one gets:

$$\frac{\partial^2 \mathbf{E}}{\partial t^2} = \Delta\mathbf{E} + [(\nabla\rho \cdot \mathbf{V})\mathbf{V} - c^2\nabla\rho] + \rho\mathbf{V} \, \mathrm{div}\mathbf{V} - \rho\frac{\partial \mathbf{V}}{\partial t} \tag{30}$$

which, for $\rho = 0$, yields the classical vector wave equation. For example, when $\mathbf{V}$ is a stationary field orthogonal to $\mathbf{B}$ with $\mathrm{div}\mathbf{V} = 0$, the operator $\Delta\mathbf{E} + [(\nabla\rho \cdot \mathbf{V})\mathbf{V} - c^2\nabla\rho]$ only contains second partial derivatives in the direction of $\mathbf{V}$, explaining why concentrated solitary waves have chances to be generated.

D   Einstein's equation

All the results presented up to this moment can be extended to general metric spaces by writing the equations in *covariant* form. To this end, it is sufficient to recall that (9), (24), (25), (26), (28), already hold for a generic metric tensor $g_{\alpha\beta}$. The Lorenz condition (8) and the continuity equation (22) take now respectively the form:

$$\partial_\alpha A^\alpha = 0 \tag{31}$$

$$\partial_\alpha(\rho V^\alpha) = 0. \tag{32}$$

We recall that the symmetric *electromagnetic stress tensor* turns out to be defined as:

$$U^{\alpha\beta} = g^{\alpha\gamma}F_{\gamma\delta}F^{\delta\beta} + \tfrac{1}{4}\,g^{\alpha\beta}F_{\mu\lambda}F^{\mu\lambda} \tag{33}$$

and that Maxwell's equations are compatible with the conservation law:

$$\partial_\alpha U^{\alpha\beta} = 0. \tag{34}$$

We show (34) for the *flat* metric tensor $g_{\alpha\beta} = \text{diag}\{1, -1, -1, -1\}$ and the Cartesian system of coordinates $(x_0, x_1, x_2, x_3) = (ct, x, y, z)$. A general proof in covariant form is given for instance in [Jackson(1975)]. One has:

$$\frac{\partial U^{0\beta}}{\partial x_\beta} = \frac{1}{2c}\frac{\partial}{\partial t}(|\mathbf{E}|^2 + c^2|\mathbf{B}|^2) + c\,\text{div}(\mathbf{E} \times \mathbf{B}) \tag{35}$$

and

$$\left(\frac{\partial U^{1\beta}}{\partial x_\beta}, \frac{\partial U^{2\beta}}{\partial x_\beta}, \frac{\partial U^{2\beta}}{\partial x_\beta}\right) = \left(\frac{\partial \mathbf{B}}{\partial t} + \text{curl}\mathbf{E}\right) \times \mathbf{E}$$

$$-\left(\frac{\partial \mathbf{E}}{\partial t} - c^2\text{curl}\mathbf{B}\right) \times \mathbf{B} + \mathbf{E}\,\text{div}\mathbf{E} + c^2\mathbf{B}\,\text{div}\mathbf{B}. \tag{36}$$

Therefore, if Maxwell's equations in vacuum are satisfied, all the above expressions are zero. In particular, requiring (35) to be zero corresponds to equation (5). In the end, one gets the implication:

$$\partial_\alpha F^{\alpha\beta} = 0 \quad \Rightarrow \quad \partial_\alpha U^{\alpha\beta} = 0. \tag{37}$$

Similarly, by adding and subtracting the term $(\text{div}\mathbf{E})\mathbf{V}$, (36) becomes:

$$\left(\frac{\partial U^{1\beta}}{\partial x_\beta}, \frac{\partial U^{2\beta}}{\partial x_\beta}, \frac{\partial U^{2\beta}}{\partial x_\beta}\right) = \left(\frac{\partial \mathbf{B}}{\partial t} + \text{curl}\mathbf{E}\right) \times \mathbf{E}$$

$$-\left(\frac{\partial \mathbf{E}}{\partial t} - c^2 \mathrm{curl}\mathbf{B} + (\mathrm{div}\mathbf{E})\mathbf{V}\right) \times \mathbf{B} + (\mathbf{E} + \mathbf{V} \times \mathbf{B})\,\mathrm{div}\mathbf{E} + c^2\mathbf{B}\,\mathrm{div}\mathbf{B}$$

which is now compatible with the new set of equations (16), (17), (18), (19). Hence, we now have the implication (see [Funaro(2008)]):

$$\partial_\alpha F^{\alpha\beta} = \frac{\rho}{c}V^\beta \quad \text{and} \quad F^{\alpha\beta}V_\beta = 0 \quad \Rightarrow \quad \partial_\alpha U^{\alpha\beta} = 0 \qquad (38)$$

which is very important since it says that (34) is true under milder hypotheses, because the space of solutions of the new set of equations is much larger than that corresponding to Maxwell's equations.

We can now introduce Einstein's equation (see, e.g.: [Fock(1959)]). For a given constant χ and a given tensor $T_{\alpha\beta}$ satisfying:

$$\partial_\alpha T^{\alpha\beta} = 0 \qquad (39)$$

the goal is to find a metric tensor $g_{\alpha\beta}$ such that:

$$G_{\alpha\beta} = -\chi T_{\alpha\beta} \qquad \text{where} \quad G_{\alpha\beta} = R_{\alpha\beta} - \tfrac{1}{2}g_{\alpha\beta}R. \qquad (40)$$

The *signature* of the metric tensor is supposed to be $(+,-,-,-)$. The constant $\chi > 0$ will be specified in appendix E. In (40), $R_{\alpha\beta}$ is the *Ricci curvature tensor* and R is the *scalar curvature*. They are defined as:

$$R_{\alpha\beta} = \frac{\partial \Gamma^\delta_{\alpha\beta}}{\partial x_\delta} - \frac{\partial \Gamma^\delta_{\alpha\delta}}{\partial x_\beta} + \Gamma^\gamma_{\alpha\beta}\Gamma^\delta_{\gamma\delta} - \Gamma^\gamma_{\alpha\delta}\Gamma^\delta_{\beta\gamma} \qquad (41)$$

$$R = g^{\alpha\beta}R_{\alpha\beta} \qquad (42)$$

where the *Christoffel symbols* are:

$$\Gamma^\delta_{\alpha\beta} = \frac{g^{\delta\gamma}}{2}\left(\frac{\partial g_{\gamma\alpha}}{\partial x_\beta} + \frac{\partial g_{\gamma\beta}}{\partial x_\alpha} - \frac{\partial g_{\alpha\beta}}{\partial x_\gamma}\right). \qquad (43)$$

One can show that the left-hand side of (40) is always compatible with (39), i.e.: $\partial_\alpha G^{\alpha\beta} = 0$, $\forall g_{\alpha\beta}$. Solutions of Einstein's equation are proposed in appendix E when the right-hand side of (40) is the electromagnetic stress tensor (i.e., $T_{\alpha\beta}$ is proportional to $U_{\alpha\beta}$).

E   Exact solutions

Let us start by working in Cartesian coordinates (x, y, z). By orienting the electric field along the z-axis, a full solution of the set of equations (16), (17), (18), (19) is obtained with the following setting:

$$\mathbf{E} = \Big(0,\ 0,\ cf(z)g(ct - x)\Big)$$

$$\mathbf{B} = \Big(0, \ -f(z)g(ct-x), \ 0\Big) \qquad \mathbf{V} = (c, \ 0, \ 0) \qquad (44)$$

where f and g can be arbitrary. This wave is shifting at the speed of light along the x-axis. Note that it is possible to enforce the condition $\rho = 0$ (thus, returning to the set of Maxwell's equations) only if f is the constant function. This first example shows why the new solution space turns out to be rather large. Regarding (44), we can also provide the potentials (see (7)):

$$\mathbf{A} = \Big(-cF(z)g(ct-x), \ 0, \ 0\Big) \qquad \Phi = -cF(z)g(ct-x) \qquad (45)$$

where F is a primitive of f. Note that (27) is satisfied.

An upgraded case is obtained as follows:

$$\mathbf{E} = \Big(0, \ cf_1(y,z)g(ct-x), \ cf_2(y,z)g(ct-x)\Big)$$

$$\mathbf{B} = \Big(0, \ -f_2(y,z)g(ct-x), \ f_1(y,z)g(ct-x)\Big) \qquad \mathbf{V} = (c, \ 0, \ 0). \quad (46)$$

In order to enforce condition (18), f_1 and f_2 must satisfy:

$$\frac{\partial f_2}{\partial y} = \frac{\partial f_1}{\partial z}. \qquad (47)$$

If f_1, f_2 and g have compact support, the wave remains bounded in a shifting portion of space. Again, we can have $\rho = 0$ only if f_1 and f_2 are both constants, drastically reducing the dimension of the solution space. As far as the potentials are concerned, we have:

$$\mathbf{A} = \Big(-cF(y,z)g(ct-x), \ 0, \ 0\Big) \qquad \Phi = -cF(y,z)g(ct-x) \qquad (48)$$

where F is such that $f_1 = \partial F/\partial y$ and $f_2 = \partial F/\partial z$. This choice is possible by virtue of (47).

Solutions similar to the above ones can be written in cylindrical coordinates (r, ϕ, z). For instance, we may consider:

$$\mathbf{E} = \Big(cf(r)g(ct-z), \ 0, \ 0\Big)$$

$$\mathbf{B} = \Big(0, \ f(r)g(ct-z), \ 0\Big) \qquad \mathbf{V} = (0, \ 0, \ c) \qquad (49)$$

$$\mathbf{A} = \Big(0, \ 0, \ -cF(r)g(ct-z)\Big) \qquad \Phi = -cF(r)g(ct-z) \qquad (50)$$

where now the shifting is in the direction of the z-axis. Again, F is a primitive of f. The function f must tend to zero for $r \to 0$, in order to be able to define the fields on the z-axis. Here, $\mathbf{E}$ is radial and $\mathbf{B}$ is organized in c_osed loops around the z-axis. Starting from this setting there are no solutioas (except zero) in the Maxwellian case. Further extensions, explicitly depending also on the variable ϕ, may be considered. If f and g have compact support, one obtains a kind of bullet travelling undisturbed at the speed cf light (see figure 1.2).

By arguing in spherical coordinates (r, θ, ϕ), one can analyze waves associated to perfect spherical fronts. We have:

$$\mathbf{E} = \left(0, \; \frac{c}{r} f_1(\theta, \phi) g(ct - r), \; \frac{c}{r} f_2(\theta, \phi) g(ct - r) \right)$$

$$\mathbf{B} = \left(0, -\frac{1}{r} f_2(\theta, \phi) g(ct - r), \frac{1}{r} f_1(\theta, \phi) g(ct - r) \right) \qquad \mathbf{V} = (c, \; 0, \; 0) \quad (51)$$

$$\mathbf{A} = \left(-cF(\theta, \phi) g(ct - r), \; 0, \; 0 \right) \qquad \Phi = -cF(\theta, \phi) g(ct - r) \qquad (52)$$

with the cond_tions:

$$\frac{\partial(f_2 \sin \vartheta)}{\partial \theta} = \frac{\partial f_1}{\partial \phi} \qquad f_1 = \frac{\partial(F \sin \theta)}{\partial \theta} \qquad f_2 = \frac{\partial F}{\partial \phi}. \qquad (53)$$

The electric and magnetic fields are tangent to spherical surfaces and shift in the radial direction. Let us observe that F and g may be arbitrary, while there are no waves of this type in the Maxwellian case ($\mathrm{div}\mathbf{E} = 0$).

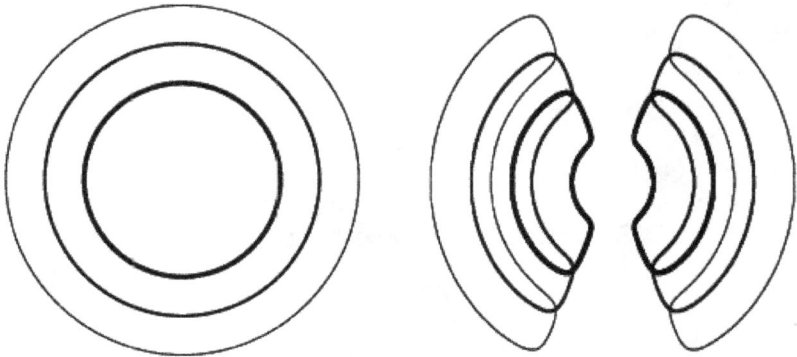

Fig. 4.3   *Qualitative sketch of the evolution of spherical wave-fronts (left), compared to the weird one of the Hertz solution (right). In the first picture, the surfaces enveloping the electromagnetic fields are spheres, whereas in the second picture they are doughnuts that self-intersect during evolution.*

Looking for spherical type solutions having $\rho = \mathrm{div}\mathbf{E} = 0$, one finds oneself with very few possibilities. An example is the *Hertzian dipole* solution:

$$\mathbf{E} = \left(\frac{2\cos\theta}{r^2} \left(\frac{g_1}{r} + kg_2 \right), \; \frac{\sin\theta}{r} \left(\frac{kg_2}{r} + (r^{-2} - k^2)g_1 \right), \; 0 \right) \tag{54}$$

deducible from:

$$\mathbf{A} = \left(\frac{g_2 k \cos\theta}{r}, \; -\frac{g_2 k \sin\theta}{r}, \; 0 \right) \qquad \Phi = \frac{k\cos\theta}{r} \left(kg_2 + \frac{g_1}{r} \right) \tag{55}$$

where $k \geq 1$ is an integer and $g_1 = \sin k(ct - r)$, $g_2 = \cos k(ct - r)$.

Other similar examples may be taken into account; however, the restriction $\rho = 0$ is too heavy to allow much freedom. For example, the function g in (51) is arbitrary and regulates the longitudinal shape of the wave. It does not interfere with f_1 and f_2, which control the transversal behavior. By expanding g in Fourier series through the basis functions $g_1 = \sin k(ct - r)$ and $g_2 = \cos k(ct - r)$, for $k \geq 1$, we can adjust accordingly the solutions given in (54) for an arbitrary k. Regarding the transversal behavior, the Hertzian solution (longitudinally modulated by g) is now "blocked" by the coefficients of the expansion, reducing drastically its degree of freedom.

The Hertz solution does not represent a free-wave, since it does not satisfy either $\mathbf{E} + \mathbf{V} \times \mathbf{B} = 0$ or (27). The envelope of the electromagnetic fields does not correspond to spheres, although their development resembles that of spherical fronts. In fact, in the Maxwellian case, one has to work with doubly connected topological objects, so that the fronts have toroid shape and do not move according to the rules of geometrical optics (see figure 4.3 and the explanations given in [Funaro(2008)]). Here the velocity field $\mathbf{V}$ is not defined and there is no reasonable way to introduce it, compatibly with both the direction of the energy transfer of the wave and the geometric advancing of the spherical fronts.

We now examine some solutions of Einstein's equation. We start by setting:

$$T^{\alpha\beta} = -\frac{\mu^2}{c^4} U^{\alpha\beta} \tag{56}$$

where $U^{\alpha\beta}$ is the electromagnetic stress tensor (see (33)) and μ is a given constant (see (89)). We then plug the tensor $T_{\alpha\beta} = g_{\alpha\gamma}g_{\beta\delta}T^{\gamma\delta}$ on the right-hand side of Einstein's equation (40). The goal is to find the metric tensor $g_{\alpha\beta}$, representing the geometrical deformation of the space-time in correspondence to the evolution of a given electromagnetic phenomenon.

Concerning free-waves, exact solutions in a very wide context have been given in [Funaro(2008)] and [Funaro(2009b)]. In particular, exact metric tensors have been computed, by substituting in (56) the electromagnetic stress tensors associated with the radiations in (44), (46), (51). For the sake of simplicity, we would just discuss the case corresponding to (49).

By setting $u = E_1 = cB_2 = cf(r)g(ct - z)$, the electromagnetic tensor (10) in cylindrical coordinates takes the form:

$$F_{\alpha\beta} = \begin{pmatrix} 0 & u & 0 & 0 \\ -u & 0 & 0 & u \\ 0 & 0 & 0 & 0 \\ 0 & -u & 0 & 0 \end{pmatrix}. \tag{57}$$

Let us observe that, for a particular orientation of the fields, the variable r does not explicitly appear in the change of coordinates from Cartesian to cylindrical (it multiplies instead other zero entries).

We look for a metric tensor of the following simple form: diag$\{1, -\sigma^2 f^2, -1, -1\}$, where σ is a function of the single variable $\xi = ct - z$. With this hypothesis, the contravariant tensor $F^{\alpha\beta}$ and the tensor $T_{\alpha\beta}$ become respectively:

$$F^{\alpha\beta} = \begin{pmatrix} 0 & -u/(\sigma f)^2 & 0 & 0 \\ u/(\sigma f)^2 & 0 & 0 & u/(\sigma f)^2 \\ 0 & 0 & 0 & 0 \\ 0 & -u/(\sigma f)^2 & 0 & 0 \end{pmatrix} \tag{58}$$

$$T_{\alpha\beta} = -\frac{\mu^2}{c^4} \begin{pmatrix} (u/\sigma f)^2 & 0 & 0 & (u/\sigma f)^2 \\ 0 & 0 & 0 & 0 \\ 0 & 0 & 0 & 0 \\ (u/\sigma f)^2 & 0 & 0 & (u/\sigma f)^2 \end{pmatrix}. \tag{59}$$

On the other hand, from $g_{\alpha\beta}$, we can first compute the Christoffel symbols (see (43)) and then the entries of the Ricci tensor (see (41) and (42)), obtaining:

$$R_{00} = R_{03} = R_{30} = R_{33} = \frac{\sigma''}{\sigma} \qquad R = 0 \tag{60}$$

where upper primes denote derivation with respect to the variable ξ. All the other entries turn out to be zero.

Substituting (59) and (60) in (40), one arrives at the differential equation:

$$-\sigma''\sigma = \frac{\chi\mu^2}{c^2}g^2 \tag{61}$$

where we recall that $u/f = cg$. There are interesting solutions of the above ordinary differential equation. For example:

$$g = \sin \omega(ct - z) \qquad \Rightarrow \qquad \sigma = \frac{\mu\sqrt{\chi}}{c\omega}\sin\omega(ct - z) \qquad (62)$$

shows that the intensity of the gravitational wave produced is inversely proportional to the frequency of the originating electromagnetic signal. Note that the divergence of the electric field, which is not zero in the flat space, is always equal to zero in the modified metric space.

It is relevant to point out that the sign of the electromagnetic tensor on the right-hand side of Einstein's equation is opposite to the one usually assumed. The explanation of this fact is given in [Funaro(2008)] and [Funaro(2009b)]. One practical reason relies on the possibility of finding solutions to equation (61). In fact, if we switch the sign of the right-hand side of (61) there are no chances of finding meaningful bounded solutions.

The change of sign of the right-hand side of Einstein's equation does not affect other consolidated solutions, that can be easily adapted to the new context. In a new version of the Reissner-Nordström metric (see [Misner, Thorne and Wheeler(1973)]), the nonzero entries can be defined as follows:

$$g_{00} = 1 - \frac{M}{r} - \frac{Q^2}{r^2}, \quad g_{11} = -1/g_{00}, \quad g_{22} = -r^2, \quad g_{33} = -(r\sin\theta)^2 \quad (63)$$

for $M > 0$ and Q related respectively to the mass and the charge of a *black-hole*.

After introducing the electromagnetic potential $A_\alpha = (A_0, 0, 0, 0)$, where $A_0 = c^2 Q/(r\mu\sqrt{\chi/2})$, one computes the corresponding tensor $U_{\alpha\beta}$ and successively finds out that (40) is satisfied with the tensor (56) on the right-hand side.

It is crucial to observe that in the standard Reissner-Nordström metric, the first entry in (63) is replaced by $g_{00} = 1 - M/r + Q^2/r^2$, which solves Einstein's equation where the plus sign appears on the right-hand side of (40). Differently from the standard case, one is able to remove the constraint $M > 2Q$ using (63), therefore allowing the existence of the so called *horizon* for small masses.

An estimate of the constant χ has been provided in [Funaro(2009b)] by assuming the horizon to have the same magnitude as the electron radius η. In this way one gets: $\chi \approx 32(\pi c^2 \eta \epsilon_0/\mu e)^2$, where e is the electron's charge. According to the computations carried out in appendix H, one finds out that the order of magnitude of the adimensional constant χ is about 0.1759.

F   Lorentz invariance

In the 4D space, we have the frame $(x_0, x_1, x_2, x_3) = (ct, x, y, z)$, considered to be at rest, and another inertial frame $(\tilde{x}_0, \tilde{x}_1, \tilde{x}_2, \tilde{x}_3)$ shifting at constant velocity $\mathbf{v} = (v, 0, 0)$, $|v| < c$. According to Lorentz, we can write:

$$(\tilde{x}_0, \tilde{x}_1, \tilde{x}_2, \tilde{x}_3) = (\gamma ct - \beta\gamma x, \ -\beta\gamma ct + \gamma x, \ y, \ z) \qquad (64)$$

where $\beta = v/c$ and $\gamma = 1/\sqrt{1 - \beta^2}$. Based on this definition, it is easy to evaluate the Jacobian matrix $a_{\alpha\beta} = \partial x_\alpha / \partial \tilde{x}_\beta$ and its inverse $a^{\alpha\beta}$.

The contravariant electromagnetic tensor in the moving frame is defined as $\tilde{F}^{\alpha\beta} = a^{\alpha\gamma} F^{\gamma\delta} a^{\delta\beta}$, where $F^{\gamma\delta}$ is given in (11). Explicitly, one has:

$$\begin{pmatrix} 0 & -E_1 & -\gamma(E_2 - vB_3) & -\gamma(E_3 + vB_2) \\[2mm] E_1 & 0 & \dfrac{\gamma}{c}(vE_2 - c^2B_3) & \dfrac{\gamma}{c}(vE_3 + c^2B_2) \\[2mm] \gamma(E_2 - vE_3) & -\dfrac{\gamma}{c}(vE_2 - c^2B_3) & 0 & -cB_1 \\[2mm] \gamma(E_3 + vB_2) & -\dfrac{\gamma}{c}(vE_3 + c^2B_2) & cB_1 & 0 \end{pmatrix}.$$

We first check the validity of Maxwell's equations in the moving frame. To this end, we must prove the counterpart of (12), i.e.:

$$\frac{\partial \tilde{F}^{\alpha\beta}}{\partial \tilde{x}_\alpha} = 0 \qquad \text{for } \beta = 0, 1, 2, 3. \qquad (65)$$

We have:

$$D_0 = \frac{\partial \tilde{F}^{\alpha 0}}{\partial \tilde{x}_\alpha} = \frac{\beta\gamma}{c}\left(\frac{\partial \mathbf{E}}{\partial t} - c^2 \mathrm{curl}\mathbf{B}\right)_1 + \gamma\,\mathrm{div}\mathbf{E} \qquad (66)$$

$$D_1 = \frac{\partial \tilde{F}^{\alpha 1}}{\partial \tilde{x}_\alpha} = -\frac{\gamma}{c}\left(\frac{\partial \mathbf{E}}{\partial t} - c^2 \mathrm{curl}\mathbf{B}\right)_1 - \beta\gamma\,\mathrm{div}\mathbf{E} \qquad (67)$$

$$D_2 = \frac{\partial \tilde{F}^{\alpha 2}}{\partial \tilde{x}_\alpha} = -\frac{1}{c}\left(\frac{\partial \mathbf{E}}{\partial t} - c^2 \mathrm{curl}\mathbf{B}\right)_2 \qquad (68)$$

$$D_3 = \frac{\partial \tilde{F}^{\alpha 3}}{\partial \tilde{x}_\alpha} = -\frac{1}{c}\left(\frac{\partial \mathbf{E}}{\partial t} - c^2 \mathrm{curl}\mathbf{B}\right)_3 \qquad (69)$$

where $(\ )_m$ denotes the m-th component of the vector. In the above expressions, the operators *curl* and *div* are associated with the coordinates

(x, y, z) of the reference frame at rest. Clearly, when (1) and (3) are satisfied, we have $D_m = 0$ for $m = 0, 1, 2, 3$. This is basically the proof of the Lorentz invariance given by A. Einstein.

In order to handle the new set of equations, we start by proving the invariance of (19). We define $\tilde{V}_\beta = (c, -\tilde{\mathbf{V}})$ and $\tilde{V}^\beta = (c, \tilde{\mathbf{V}})$, where in the moving frame $\tilde{\mathbf{V}}$ is the velocity field corresponding to $\mathbf{V}$. Note that, according to the theory of relativity, the expression $\tilde{\mathbf{V}} = \mathbf{V} - \mathbf{v}$ is incorrect. We assume instead that:

$$\tilde{\mathbf{V}} = \frac{1}{1 - vV_1/c^2} \left(V_1 - v, \frac{V_2}{\gamma}, \frac{V_3}{\gamma} \right) \tag{70}$$

which is exactly the rule to sum up velocities in special relativity. We are going to show that this choice is in agreement with the invariance properties of all the equations. Thus, first of all we must have:

$$\tilde{F}^{\alpha\beta} \tilde{V}_\beta = 0 \quad \text{for } \beta = 0, 1, 2, 3. \tag{71}$$

By substituting the expression (70) in (71), one obtains:

$$\tilde{F}^{0\beta} \tilde{V}_\beta = \frac{1}{1 - vV_1/c^2} \left(-v(\mathbf{E} + \mathbf{V} \times \mathbf{B})_1 + \mathbf{E} \cdot \mathbf{V} \right) \tag{72}$$

$$\tilde{F}^{1\beta} \tilde{V}_\beta = \frac{1}{1 - vV_1/c^2} \left(c(\mathbf{E} + \mathbf{V} \times \mathbf{B})_1 - \beta\, \mathbf{E} \cdot \mathbf{V} \right) \tag{73}$$

$$\tilde{F}^{2\beta} \tilde{V}_\beta = \frac{c}{\gamma(1 - vV_1/c^2)} (\mathbf{E} + \mathbf{V} \times \mathbf{B})_2 \tag{74}$$

$$\tilde{F}^{3\beta} \tilde{V}_\beta = \frac{c}{\gamma(1 - vV_1/c^2)} (\mathbf{E} + \mathbf{V} \times \mathbf{B})_3. \tag{75}$$

Considering that $\mathbf{E}$ is orthogonal to $\mathbf{V}$ and recalling (19), all the above expressions vanish. This means that (71) is true.

Regarding the invariance of the differential equation (16), we would like to have (see (23)):

$$\frac{\partial \tilde{F}^{\alpha\beta}}{\partial \tilde{x}_\alpha} - \frac{1}{c} \frac{\partial \tilde{F}^{\gamma 0}}{\partial \tilde{x}_\gamma} \tilde{V}^\beta = 0 \quad \text{for } \beta = 1, 2, 3. \tag{76}$$

Going back to (66), we rewrite D_0 as follows:

$$D_0 = \frac{\beta\gamma}{c} \left(\frac{\partial \mathbf{E}}{\partial t} - c^2 \mathrm{curl}\mathbf{B} + \mathbf{V} \mathrm{div}\mathbf{E} \right)_1 + \gamma \left(1 - \frac{vV_1}{c^2} \right) \mathrm{div}\mathbf{E}. \tag{77}$$

By assuming that (16) is satisfied, then the first term on the right-hand side of (77) is zero. After this simplification one gets:

$$D_0 = \gamma \left(1 - \frac{vV_1}{c^2} \right) \mathrm{div}\mathbf{E}. \tag{78}$$

Similarly, from (67) we have:

$$D_1 = -\frac{\gamma}{c}\left(\frac{\partial \mathbf{E}}{\partial t} - c^2\text{curl}\mathbf{B} + \mathbf{V}\,\text{div}\mathbf{E}\right)_1$$

$$+ \gamma\left(\frac{V_1}{c} - \beta\right)\text{div}\mathbf{E} = \frac{\gamma}{c}(V_1 - v)\,\text{div}\mathbf{E}. \tag{79}$$

Recalling that $\tilde{V}^1 = (V_1 - v)/(1 - vV_1/c^2)$ (see (70)), we get:

$$\frac{\partial \tilde{F}^{\alpha 1}}{\partial \tilde{x}_\alpha} - \frac{\tilde{V}^1}{c}\frac{\partial \tilde{F}^{\gamma 0}}{\partial \tilde{x}_\gamma} = D_1 - \frac{V_1 - v}{c(1 - vV_1/c^2)}D_0 = 0. \tag{80}$$

The last expression is zero thanks to (78) and (79). Thus, we have obtained (76) for $\beta = 1$. With the same arguments, for $\beta = 2$ and $\beta = 3$, we have:

$$\frac{\partial \tilde{F}^{\beta\alpha}}{\partial \tilde{x}_\alpha} - \frac{\tilde{V}^\beta}{c}\frac{\partial \tilde{F}^{0\gamma}}{\partial \tilde{x}_\gamma} = D_\beta - \frac{V_\beta}{c\gamma(1 - vV_1/c^2)}D_0$$

$$= -\frac{1}{c}\left(\frac{\partial \mathbf{E}}{\partial t} - c^2\text{curl}\mathbf{B}\right)_\beta - \frac{V_\beta}{c}\,\text{div}\mathbf{E}$$

$$= -\frac{1}{c}\left(\frac{\partial \mathbf{E}}{\partial t} - c^2\text{curl}\mathbf{B} + \mathbf{V}\,\text{div}\mathbf{E}\right)_\beta \tag{81}$$

that due to (16) is also zero. Hence, (76) is proven. Because of the numerous constraints, we do not expect there are other interesting choices for $\tilde{\mathbf{V}}$, except for the one given in (70).

We would like to note that:

$$|\tilde{\mathbf{V}}| = c \tag{82}$$

that corresponds to (20). In this way light moves at the same speed in both the rest frame and the one in motion.

Finally, it is possible to rewrite some exact solution (for example (44)) in the moving frame, obtaining:

$$\tilde{\mathbf{E}} = \left(0,\ 0,\ c\gamma(1-\beta)f(\tilde{z})g(\gamma(1-\beta)(c\tilde{t} - \tilde{x}))\right)$$

$$\tilde{\mathbf{B}} = \left(0,\ -\gamma(1-\beta)f(\tilde{z})g(\gamma(1-\beta)(c\tilde{t} - \tilde{x})),\ 0\right) \quad \tilde{\mathbf{V}} = (c,\ 0,\ 0). \tag{83}$$

We observe that the wave undergoes contraction or expansion in the direction of motion depending on the sign of β. This is also true if g corresponds to a single pulse, thus in that case there is no continuous oscillatory phenomenon directly involved. Note that the energy changes with the observer.

G The general set of model equations

We are ready to introduce the final set of model equations, which strictly includes the case of free-waves previously discussed. It is just a matter of transforming equation (19) in order to include situations where the fronts may develop without respecting the rules of geometrical optics in the flat space. We have:

$$\frac{\partial \mathbf{E}}{\partial t} = c^2 \mathrm{curl} \mathbf{B} - \rho \mathbf{V} \tag{84}$$

$$\frac{\partial \mathbf{B}}{\partial t} = -\mathrm{curl} \mathbf{E} \tag{85}$$

$$\mathrm{div} \mathbf{B} = 0 \tag{86}$$

$$\rho \left(\frac{D\mathbf{V}}{Dt} + \mu(\mathbf{E} + \mathbf{V} \times \mathbf{B}) \right) = -\nabla p \tag{87}$$

with $\rho = \mathrm{div}\mathbf{E}$ and:

$$\frac{D\mathbf{V}}{Dt} = \frac{\partial \mathbf{V}}{\partial t} + (\mathbf{V} \cdot \nabla)\mathbf{V}. \tag{88}$$

Relation (87) recalls the Euler's equation for inviscid fluids with a forcing term of electromagnetic type given by the vector: $\mathbf{E} + \mathbf{V} \times \mathbf{B}$. The Maxwell's case is now obtained by setting $\rho = 0$ and $p = 0$. It should be also clear that when $D\mathbf{V}/Dt = 0$ and $p = 0$, one comes back to the case of free-waves.

Up to dimensional scaling, the scalar p plays the role of pressure. The constant μ is a charge divided by a mass and in appendix H it is estimated to be equal to:

$$\mu \approx 2.85 \times 10^{11} \ \mathrm{Coulomb/kg} \tag{89}$$

which is of the same order of magnitude as the elementary charge divided by the electron's mass.

A further equation, related to energy conservation arguments (see later), can be added:

$$\frac{\partial p}{\partial t} = \mu \rho \mathbf{E} \cdot \mathbf{V}. \tag{90}$$

This says that pressure may rise as a consequence of a lack of orthogonality between $\mathbf{E}$ and $\mathbf{V}$. These two vectors are instead always orthogonal in the case of free-waves (multiply (19) by $\mathbf{V}$).

By multiplying scalarly (87) by $\mathbf{V}$ and taking into account of (90), we arrive at a *Bernoulli* type equation:

$$\frac{\rho}{2}\frac{D|\mathbf{V}|^2}{Dt} + \frac{Dp}{Dt} = 0. \tag{91}$$

The writing of the above model equations in covariant form read as follows:

$$\partial_\alpha F^{\alpha\beta} = \frac{\rho}{c} V^\beta \tag{92}$$

$$\rho\left(V^\gamma \partial_\gamma V^\alpha + \frac{\mu}{c}F^{\alpha\beta}V_\beta\right) = \partial^\alpha p. \tag{93}$$

By setting $V^0 = c$, we actually get (84) from (92) with $\beta = 1, 2, 3$ (for $\beta = 0$ such a relation is just a trivial identity). We then get (87) from (93) with $\beta = 1, 2, 3$, and (90) from (93) with $\beta = 0$. The continuity equation (32) naturally descends from taking the divergence of (92).

We now define:

$$T_{\alpha\beta} = \frac{\mu}{c^4}\left(-\mu U_{\alpha\beta} + M_{\alpha\beta}\right) \tag{94}$$

that generalizes (56). In (94), $M_{\alpha\beta}$ is a *mass* type tensor, defined as follows:

$$M_{\alpha\beta} = \rho V_\alpha V_\beta - p\, g_{\alpha\beta}. \tag{95}$$

In addition, we assume the condition:

$$g^{\alpha\beta}V_\alpha V_\beta = 0 \tag{96}$$

that corresponds to the eikonal equation in a generic metric space, extending the validity to (20).

One can show that, if (92) and (93) are true then (39) also holds true. This is the way to show the compatibility of the model equations with energy preservation properties. For the sake of simplicity we omit the computation (see [Funaro(2009b)] for details).

By taking the *trace* of (40), we get this interesting relation between pressure and scalar curvature:

$$p = \frac{c^4}{4\chi\mu}R. \tag{97}$$

We can finally give a definition of *mass density* ρ_m. When $g_{\alpha\beta} = \mathrm{diag}\{1, -1, -1, -1\}$ and $|V_0| = c$, we set:

$$\rho_m = \frac{\epsilon_0 M_{00}}{\mu c^2} = \frac{\epsilon_0}{\mu}\left(\rho - \frac{p}{c^2}\right). \tag{98}$$

By integrating ρ_m on the "support" Ω of the wave we get its "mass" m at rest. For instance, a Maxwellian wave has $\rho = 0$ and $p = 0$, so that $m = 0$, as expected. For a free-wave, one has $p = 0$ and $\rho \neq 0$. However, it is true that having $\mathbf{E}$ with compact support implies: $\int_{\Omega} \rho = \int_{\partial\Omega} \mathbf{E} = 0$, so that once again one obtains $m = 0$. Several other solutions may instead produce nonzero masses. We are going to discuss an example soon (see also appendix H).

Following [Funaro(2008)], we work in the left-handed cylindrical system of coordinates (r, z, ϕ) (the right-handed version is (r, ϕ, z)). For an integer $k \geq 1$ and arbitrary constants $\omega > 0$, γ_0 and γ_1, we start by defining the potentials:

$$\mathbf{A} = -\left(\frac{\gamma_1}{\omega} J_{k+1}(\omega r) \sin(c\omega t - k\phi), \quad 0, \quad \frac{\gamma_0 r^3}{4k} + \frac{\gamma_1}{\omega} J_{k+1}(\omega r) \cos(c\omega t - k\phi)\right)$$

$$\Phi = -\frac{\gamma_0 r^2}{2\omega} - \frac{\gamma_1}{\omega} J_k(\omega r) \cos(c\omega t - k\phi) \tag{99}$$

for $0 \leq \phi < 2\pi$, $0 \leq r \leq \delta_k/\omega$ and any z. By J_k we denoted the k-th *Bessel function*, recoverable from the relation:

$$J_k''(x) + \frac{J_k'(x)}{x} - \frac{k^2 J_k(x)}{x^2} + J_k(x) = 0. \tag{100}$$

The quantity δ_k is the first zero of J_k. Dimensionally, ω is the inverse of a length.

The potentials satisfy the Lorenz gauge condition (8) and bring, thanks to (7), to the electromagnetic fields:

$$\mathbf{E} = \left(\frac{\gamma_0 r}{\omega} + \gamma_1 \frac{k J_k(\omega r)}{\omega r} \cos(c\omega t - k\phi), \quad 0, \quad \gamma_1 J_k'(\omega r) \sin(c\omega t - k\phi)\right)$$

$$\mathbf{B} = \frac{1}{c}\left(0, \quad \frac{\gamma_0 r^2}{k} + \gamma_1 J_k(\omega r) \cos(c\omega t - k\phi), \quad 0\right). \tag{101}$$

In order to check the above expressions it is worthwhile recalling the relations: $J_{k+1}'(x) + (k+1)J_{k+1}(x)/x = J_k(x)$ and $J_k'(x) - kJ_k(x)/x = -J_{k+1}(x)$. For $r = 0$, we have $\mathbf{B} = 0$ and $\mathbf{E} = 0$, if $k > 1$. In the case $k = 1$, we have $\mathbf{B} = 0$ and $\mathbf{E} = \gamma_1 J_1'(0)(\cos(c\omega t - \phi), 0, \sin(c\omega t - \phi))$. For $t = 0$, the last expression translated in Cartesian coordinates is: $\gamma_1 J_1'(0)(1, 0, 0)$ (see figure 4.4).

As far as the pressure is concerned, one has:

$$p = p_0 + \frac{c^2 \omega \gamma_0 r^2}{k^2} - \frac{\mu \gamma_0^2 r^2}{\omega^2}\left(1 - \frac{\omega^2 r^2}{2k^2}\right)$$

$$- \frac{2\mu \gamma_0 \gamma_1 r}{k\omega} J_k'(\omega r) \cos(c\omega t - k\phi) \tag{102}$$

where p_0 is an arbitrary constant.

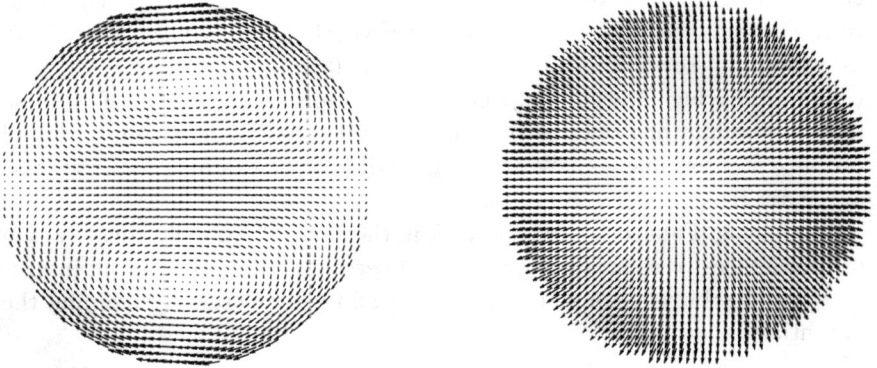

Fig. 4.4 *Displacement of the electric field, at a given time t, as prescribed by (101) for $k = 1$. The magnetic field is orthogonal to the page. The solution is defined on a cylinder. It does not depend on z and rotates about the axis. On the left, one can see the case corresponding to a pure Maxwellian wave ($\gamma_0 = 0$). Here the field V does not need to be introduced. On the right, we have a full solution of the model equations, where a stationary radial electric field ($\rho > 0$) has been added to the time dependent wave. Now V can be defined as suggested in figure 1.7.*

We finally provide the expression of the velocity vector field:

$$\mathbf{V} = \left(0, \ 0, \ \frac{c\omega r}{k}\right). \tag{103}$$

Going through tedious computations, one can check that the above defined $\mathbf{E}$, $\mathbf{B}$, $\mathbf{V}$ and p solve the entire set of model equations. In addition, the following meaningful relation can be checked:

$$c^2 \mathbf{B} \ = \ \mathbf{E} \times \mathbf{V}. \tag{104}$$

The solution is defined on a cylinder Σ (see figure 4.4) and rotates about its axis. At the boundary of Σ (i.e., when $r = \delta_k/\omega$) the time-dependent components E_1 and B_2 are zero. Along the z-axis, both $\mathbf{E}$ and $\mathbf{B}$ vanish. The divergence $\mathrm{div}\mathbf{E} = 2\gamma_0/\omega$ is constant and different from zero, except for $\gamma_0 = 0$ where we have $\rho = 0$ and $\nabla p = 0$, such that a pure Maxwellian wave is obtained.

Similar solutions may be defined on tori and other, more or less complex, toroid shaped domains also in the case $\rho = 0$ (see [Chinosi, Della Croce and Funaro(2010)]). This shows how big can be the solution space of Maxwell's equations when one is not looking for *propagating* type wave-fronts.

A further generalization, in which the wave twists in the direction of the z-axis, is obtained as follows:

$$\mathbf{E} = \left(\frac{k J_k(\zeta r)}{\zeta r} \cos(c\omega t - k\phi - \lambda z), \; 0, \; J'_k(\zeta r) \sin(c\omega t - k\phi - \lambda z) \right)$$

$$\mathbf{B} = \frac{1}{c\omega} \left(\lambda J'_k(\zeta r) \sin(c\omega t - k\phi - \lambda z), \; \zeta J_k(\zeta r) \cos(c\omega t - k\phi - \lambda z), \right.$$

$$\left. - \frac{\lambda k J_k(\zeta r)}{\zeta r} \cos(c\omega t - k\phi - \lambda z) \right) \tag{105}$$

where $\zeta = \sqrt{\omega^2 - \lambda^2}$ and λ (with $|\lambda| < \omega$) is a parameter. For $\lambda = 0$ we return to the previously examined solution.

Another very special solution in Cartesian coordinates is:

$$\mathbf{E} = \left(0, \; 0, \; c \sin \omega(ct - z) \right) \qquad \mathbf{B} = 0 \qquad \mathbf{V} = (0, 0, c) \tag{106}$$

where the electric field is perfectly longitudinal. It is easy to check that such a configuration solves the entire set of equations with the pressure term equals to $p = -\frac{1}{2}\mu c^2 [\sin \omega(ct - z)]^2$. In fact, with $\mathbf{B} = 0$ and $D\mathbf{V}/Dt = 0$, we just have to solve $\partial \mathbf{E}/\partial t = \rho \mathbf{V}$ and $\mu \rho \mathbf{E} = -\nabla p$. Exact solutions of the Einstein's equation can also be proposed in this case (see [Funaro(2008)], section 5.1).

H Construction of the electron

Solutions on annular domains can be built with size, mass and charge equal to those of the electron. The toroid is denoted by Σ and it is assumed to be adequately thin in order to be approximated with the cylindrical solutions proposed in appendix G.

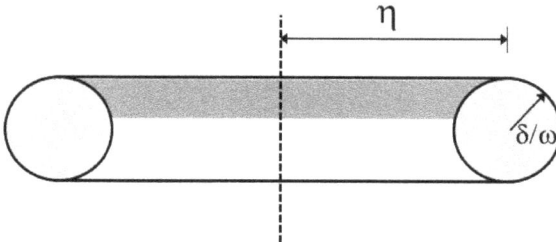

Let us take $k = 1$ in (101) and (102) (see also [Funaro(2009b)], where, having set $k = 2$, the computations are slightly different from the ones we are going to carry out here). Let $\delta = \delta_1 \approx 3.832$ be the first zero of the Bessel function J_1. After having fixed the parameter ω, the minor radius turns out to be equal to δ/ω. The major radius will be denoted by η. Thus, the volume of Σ is $2\pi^2 \eta \delta^2 / \omega^2$.

Successively, one obtains the charge $-e$ of the electron by integrating $\mathrm{div}\mathbf{E} = 2\gamma_0/\omega$ in Σ:

$$-e = \epsilon_0 \int_\Sigma \frac{2\gamma_0}{\omega} = \frac{4\pi^2 \epsilon_0 \eta \delta^2}{\omega^3} \gamma_0 \tag{107}$$

from which we can explicitly compute γ_0:

$$\gamma_0 = -\frac{e\omega^3}{4\pi^2 \epsilon_0 \eta \delta^2} = -\frac{\omega^3 \alpha h c}{2\pi^2 \eta \delta^2 e} \tag{108}$$

where $\alpha = e^2/2hc\epsilon_0$ is the *fine structure* constant and h is the Planck constant.

Let us observe that the system of coordinates (r, z, ϕ) is left-handed. Therefore, according to section 2.3, the electron has negative sign. On the contrary, working in right-handed frameworks is equivalent to dealing with antimatter, thus in this case the particle ends up to be a positron.

By taking $r = 0$ in (101), one gets $\mathbf{B} = 0$. From (103) one has $D\mathbf{V}/Dt = (-c^2\omega^2 r, 0, 0) = 0$. This means that equation (87) becomes:

$$\mu\rho\mathbf{E} = -\nabla p. \tag{109}$$

Therefore, the gradient of p is lined up with the electric field and, for symmetry reasons (see figure 4.4), one should have $p = 0$ at $r = 0$. Hence, $p_0 = 0$ in (102). The stationary part of the pressure now takes the form:

$$p_{stat} = c^2 \omega \gamma_0 r^2 - \frac{\mu \gamma_0^2 r^2}{\omega^2}\left(1 - \frac{\omega^2 r^2}{2}\right). \tag{110}$$

The mass of the particle is obtained by integrating the density ρ_m introduced in (98):

$$m = \frac{\epsilon_0}{\mu} \int_\Sigma \left(\rho - \frac{p}{c^2}\right) = 4\pi^2 \eta \frac{\epsilon_0}{\mu} \int_0^{\delta/\omega} \left(\rho - \frac{p}{c^2}\right) r\,dr. \tag{111}$$

Considering that the time-dependent part of p has zero average in Σ, going through the computations, one arrives at:

$$m = \left(\frac{\delta^2}{4} - 1\right)\frac{e}{\mu} - \left(\frac{\delta^2}{3} - 1\right)\frac{\alpha h}{8\pi^2 \eta c}. \tag{112}$$

Here γ_0 has been eliminated by recalling (108).

The next step is to set equal to zero the component of the gradient of pressure normal to the boundary of Σ. Therefore, we must have $\partial p/\partial r = 0$ for $r = \delta/\omega$. By using (87), (101) and observing that $J_1(\delta) = 0$, we get:

$$- c^2 + \frac{\mu \gamma_0}{\omega^3}(1 - \delta^2) = 0. \tag{113}$$

Recovering γ_0 from (107), the above relation implies:

$$\eta = \frac{\mu}{e} \frac{\alpha h}{2\pi^2 c} \left(1 - \frac{1}{\delta^2}\right). \tag{114}$$

Therefore, going back to (112) one computes the mass:

$$m = \frac{\delta^4 - 6\delta^2 + 6}{6(\delta^2 - 1)} \frac{e}{\mu} \approx 1.626 \frac{e}{\mu}. \tag{115}$$

Since m and e are known, we may in this way estimate the constant $\mu \approx 1.626 \, e/m \approx 2.85 \times 10^{11}$ Coulomb/kg. With this, recalling (114) we may also estimate $\eta \approx 1.35 \times 10^{-15}$ meters. The "classical electron radius" is about twice this value (see, e.g., [Haken and Wolf(1994)], p.68). Note however that multiplicative constants are systematically neglected or adjusted in the standard literature. With the help of these data we can review the estimate of the constant χ in (40), obtaining $\chi \approx 0.1759$ (see at the end of appendix E). These results are independent of the parameter ω, so that there is a family of solutions presenting the same physical properties of the electron.

We finally compute the pressure on the surface of Σ (i.e. for $r = \delta/\omega$). Neglecting the time-dependent part, from (110) we have:

$$p_{stat}(\delta/\omega) = -\frac{c^4}{\mu} \frac{\delta^4}{2(\delta^2 - 1)^2} \omega^2 = -\sigma\omega^2 < 0 \tag{116}$$

with $\sigma \approx 2.86 \times 10^{22}$. Hence, the boundary pressure behaves as ω^2.

We can finally evaluate the stationary electromagnetic energy:

$$\mathcal{E} = \frac{\epsilon_0}{2} \int_\Sigma (|\mathbf{E}|^2 + |c\mathbf{B}|^2) = \frac{3\delta^2(1 + 2\delta^2/3)}{4(\delta^4 - 6\delta^2 + 6)} mc^2 \approx 0.89 \, mc^2 \tag{117}$$

which is a result reasonably compatible with predictions.

The example here discussed does not specifically use the toroid structure of Σ. A finer study should take into account the evolution of p in an annular domain, considering that there must be a difference of pressure between the side facing the hole and the one facing outwards. Such an analysis, though far more complex, will surely reveal more interesting properties. For instance, based on the computations here developed, there are no elements for estimating the magnitude of the constant γ_1 in (101), which remains arbitrary. One could recover γ_1 by requiring the integral of the electromagnetic energy density $\frac{1}{2}\epsilon_0(|\mathbf{E}|^2 + c^2|\mathbf{B}|^2)$ of the time-dependent part to be equal to $h\nu$, where h is the Planck constant and $\nu = c\omega/2\pi$ is the associated frequency. This imposition looks at the moment artificial. Conversely, if one were able to independently fix γ_1 in relation to geometrical reasonings, this would indicate an intriguing way to estimate the Planck constant.

I   A qualitative model for the Hydrogen atom

The first step is to build a spherical magnetic background around a proton with the help of Maxwell's equations. To this end, we recall the wave equation for the magnetic field in (6). We look for solutions in spherical coordinates (r, θ, ϕ) of the type $\mathbf{B} = (0, 0, u \sin\theta)$, where u is a function only of the variables t and r. In this way, it is implicitly assumed that the spin axis of the particles is lined up with the z-axis. Under this circumstance, the wave equation is simplified as follows:

$$\frac{\partial^2 u}{\partial t^2} = c^2 \left(\frac{\partial^2 u}{\partial r^2} + \frac{2}{r}\frac{\partial u}{\partial r} - \frac{2u}{r^2} \right). \tag{118}$$

We now assume that the oscillations of $\mathbf{B}$ display a frequency ν that decays as the inverse of the distance from the origin. In practice, for a given parameter $\beta > 0$, we set:

$$\nu \approx \frac{c\beta}{2\pi r}. \tag{119}$$

Thus, arguing for example with the mode $\cos(2\pi\nu t)$, we may write:

$$\frac{\partial^2 u}{\partial t^2} \approx -(2\pi\nu)^2 u \approx -\frac{c^2\beta^2}{r^2}u. \tag{120}$$

This suggests to replace u in (118) by $\hat{u}$, where $\hat{u}$ only depends on r and satisfies the stationary equation:

$$\hat{u}'' + \frac{2}{r}\hat{u}' + \frac{\beta^2 - 2}{r^2}\hat{u} = 0. \tag{121}$$

Up to multiplicative constants, the above ordinary differential equation admits a solution of the following form:

$$\hat{u}(r) = \frac{1}{\sqrt{r}} \sin\left(\sqrt{\beta^2 - \tfrac{9}{4}} \, \log\frac{r}{\delta} \right) \qquad \text{for } \beta > \tfrac{3}{2} \tag{122}$$

with $r \geq \delta$, where $\delta > 0$ is given. In this way, we are imposing $\hat{u}(\delta) = 0$, which means that the vector $(0, 0, \hat{u}\sin\theta)$ vanishes on the surface of the sphere S_δ of radius δ, as well as along the spin axis. It is possible to prove that (121) admits only the solution zero when β is smaller than a certain amount. Note that, the substitution $r \to re^\gamma$ with $\gamma = \pi/\sqrt{\beta^2 - \frac{9}{4}}$, brings back to $\hat{u}$, up to a multiplicative constant. This implies that $\hat{u}$ is self-similar. The function $\hat{u}$ has infinite zeros and $\lim_{r \to +\infty} \hat{u}(r) = 0$. The computation of the zeros of (122) is straightforward:

$$r_k = \delta e^{\gamma k} \qquad k \geq 0. \tag{123}$$

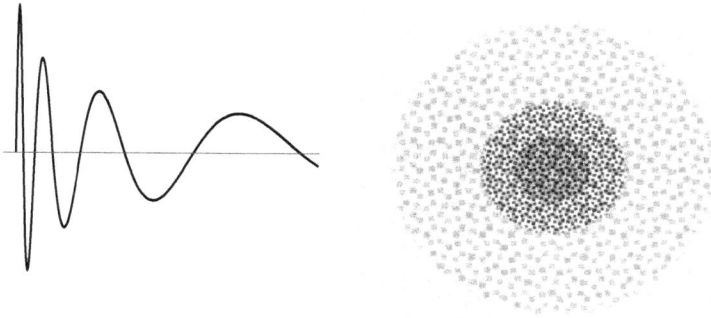

Fig. 4.5 *Qualitative behavior of the function in (122). This corresponds to a set of encapsulated shells encircling the proton.*

Thus, the inner sphere S_δ is surrounded by an infinite set of spherical shells S_k, with $S_0 = S_\delta$, whose amplitudes grow geometrically. We can assign a frequency to each shell. Inspired by (119), we can act as follows:

$$\nu_k = \frac{c\beta}{2\pi \, r_{k+1}} \qquad \forall \, r \in]r_k, r_{k+1}]. \tag{124}$$

We furnish each shell with a suitable energy $\mathcal{E}_k = h\nu_k$, h being the Planck constant. The procedure adopted so far is aimed to determine the frontiers of the encapsulated shells; however, it does not specify how the magnetic field actually behaves inside them. Nevertheless, interesting quantitative outcomes are obtained by the following arguments. If a radial stationary electric field $\mathbf{E}_0$ of intensity $I_0 = e/(4\pi\epsilon_0 r^2)$ emanates from a particle of charge e (a proton, in particular), we can compute its energy $\mathcal{E}_k$ on the domain $D_k = \mathcal{S}_{k+1} - S_k$, according to the next formula, where the integral is evaluated in spherical coordinates:

$$\mathcal{E}_k = \frac{\epsilon_0}{2} \int_{D_k} I_0^2 \, dV = 2\pi\epsilon_0 \int_{r_k}^{r_{k+1}} \left(\frac{e}{4\pi\epsilon_0 r^2}\right)^2 r^2 dr$$

$$= \frac{e^2}{8\pi\epsilon_0} \left(\frac{1}{r_k} - \frac{1}{r_{k+1}}\right) = \frac{hc\alpha(e^\gamma - 1)}{4\pi \, r_{k+1}} = \frac{\alpha(e^\gamma - 1)}{2\beta} h\nu_k. \tag{125}$$

In (125), $\alpha = e^2/2hc\epsilon_0$ is the *fine structure* constant. In the computation above, we used (123) and (124).

We actually get the Planck relation $\mathcal{E}_k = h\nu_k$, by imposing that:

$$\frac{1}{2}(e^\gamma - 1)\frac{1}{\beta} = \frac{1}{2}(e^\gamma - 1)\left(\frac{\pi^2}{\gamma^2} + \frac{9}{4}\right)^{-1/2} = \frac{1}{\alpha} \approx 137. \tag{126}$$

The equation (126) is satisfied when $\gamma \approx 6.08$, that leads us to $e^\gamma \approx 437$ and $\beta \approx 1.82$. This result is very interesting. Indeed, assuming that the proton radius is approximately $\delta = .87 \times 10^{-15}m$, we get a first shell of radius $\delta e^\gamma \approx 3.8 \times 10^{-13}m$, and a second shell of radius $\delta e^{2\gamma} \approx 1.6 \times 10^{-10}m$, which is of the same order of magnitude as the radius of the Hydrogen atom. The third shell, and the successive ones, are far larger (and energetically weaker).

It is possible to compute the global energy through the summation:

$$\mathcal{E} = \sum_{k=0}^{\infty} \mathcal{E}_k = \sum_{k=0}^{\infty} \frac{hc\beta}{2\pi \, r_{k+1}} = \frac{hc\beta}{2\pi\delta e^\gamma} \sum_{k=0}^{\infty} e^{-\gamma k}$$

$$= \frac{hc\beta}{2\pi\delta(e^\gamma - 1)} = \frac{hc\alpha}{4\pi\delta} = \frac{e^2}{8\pi\delta\epsilon_0} < \frac{3e^2}{20\pi\delta\epsilon_0}. \tag{127}$$

The very last term in (127) is the electrostatic energy of a charged ball of radius δ, as obtained by evaluating the exact integral. The discrepancy is negligible. This match is only achievable with a geometric growth of the diameters of the shells.

For an ensemble of fixed particles, a combination of the wave equation (118) together with condition (120) suggest to solve the stationary problem in vector form:

$$-\Delta\hat{\mathbf{B}} = f\,\hat{\mathbf{B}} \tag{128}$$

where f is a given function that behaves proportionally to $1/r^2$ in proximity of each particle (r being the distance from the particle). The new equation is similar to the vector version of the stationary Klein-Gordon equation with Coulomb potential.

The field $\hat{\mathbf{B}}$ is required to be zero on the surfaces of the spheres of radius $\delta > 0$ delimiting the skin of each particle, whereas curl$\hat{\mathbf{B}}$ determines the orientation of the spin axis. Note that $\hat{\mathbf{B}}$ has no direct relation with the magnetic field $\mathbf{B}$. The idea is that the border of the transition shells around the set of particles is determined by the zeros of $\hat{\mathbf{B}}$, which is exactly what has been previously done in the case of the single proton. In such a general case, the problem is of course far more complicated.

It is possible to try some computations in the case of the Hydrogen atom, i.e., when the ensemble is constituted by a proton and an electron with parallel spin axes. Working in cylindrical coordinates (r, ϕ, z), one looks for solutions of the type $\hat{\mathbf{B}} = (0, \hat{u}, 0)$, where $\hat{u}$ does not depend on the variable ϕ. The electron (with radius δ) is placed at the origin $z = 0$, whereas the proton (also of radius δ) is put at a distance d. Thus, along the segment joining the two particles, the function f takes the form:

$$f(z) = \frac{\beta_e^2}{z^2} + \frac{\beta_p^2}{(d - z)^2} \tag{129}$$

where $\beta_e > 0$ and $\beta_p > 0$ are given constants. For the proton one can set $\beta_p = \beta \approx 1.82$, as evaluated before, while it is suggested to take β_e much smaller. Substituting into (128), yields:

$$\frac{\partial^2 \hat{u}}{\partial r^2} + \frac{\partial^2 \hat{u}}{\partial z^2} + \frac{1}{r}\frac{\partial \hat{u}}{\partial r} - \frac{\hat{u}}{r^2} + f\hat{u} = 0. \tag{130}$$

Furthermore, the variables can be separated by setting $\hat{u}(r, z) = p(r)q(z)$. For a certain $\lambda > 0$, we arrive at the two equations:

$$p'' + \frac{p'}{r} - \frac{p}{r^2} - \lambda^2 p = 0 \tag{131}$$

$$q'' + f(z)q + \lambda^2 q = 0. \tag{132}$$

We study these two problems by working on a very thin cylinder connecting the electron and the proton. Up to a multiplicative constant, p behaves as r when approaching the z-axis: $p(r) \approx r + \lambda^2 r^3/8 + \lambda^4 r^5/192 + \cdots$. Concerning equation (132), the boundary conditions require that: $q(\delta) = q(d - \delta) = 0$. This is the same equation proposed in [Funaro(2009a)] in the study of the Casimir effect.

Further approximation can be introduced by observing that at a point $\bar{\delta}$ $(0 < \delta < \bar{\delta} \ll d)$ far away from the proton, one has: $f(z) \approx \beta_e^2/z^2$. If $d - \bar{\delta}$ corresponds to the radius of the second shell encircling the proton (the one having diameter of the order of $10^{-10}m$), we end up with the eigenvalue problem:

$$q'' + \left(\frac{\beta_e^2}{z^2} - \lambda^2\right)q = 0 \qquad \text{with } q(\delta) = q(\bar{\delta}) = 0. \tag{133}$$

The solution q can be explicitly expressed in terms of Bessel functions Y_α of the second kind. Indeed, up to a multiplicative constant, one has: $q(z) = Y_\alpha(\lambda z)$, with $\alpha = \sqrt{1/4 - \beta_e^2}$. If β_e is relatively small ($\beta_e \ll 1/2$), up to a multiplicative constant, one roughly gets: $q(z) \approx \cos(\lambda z)/\sqrt{z}$ (i.e., the Bessel function of the second kind with index equal to $1/2$). By imposing boundary conditions, one finally arrives at: $\lambda\delta = \pi/2$ and $\lambda\bar{\delta} = 3\pi/2$. This means that a nontrivial solution q can be only found if $\bar{\delta} = 3\delta$. Note that the result is independent of λ. This shows that, when the electron is moved sufficiently far away from the external bubble of the proton, a new shell (shaped as a ring) grows in between. This ring is assimilated to a trapped photon and can be generated or reabsorbed according to quantized energies, as the differential problem suggests. Under favorable circumstances the photon can be liberated, becoming a classical free-wave emission (see figure 3.4). At this point, in order to have quantitative confirmations, more serious discretizations of the original partial differential equation (130) are necessary.

These last disquisitions are aimed to demonstrate that quantization does not derive from innovative adjustments of the model equations, but it is hidden in the geometrical displacement of the concurring particles. Indeed, equation (128) should not be considered as an eigenvalue problem. The function f is uniquely determined according to the particles distribution. A nontrivial $\hat{\mathbf{B}}$ only exists if the reciprocal distances are correct. An illustrative example in one dimension is the boundary value problem, $-w'' = w$ with $w(a) = w(b) = 0$, in which nonvanishing solutions are permitted only when $b - a = k\pi$, with k as an integer. This means that the distance between the points a and b must attain quantized values.

Bibliography

Non sono mai uscito dai libri:
lo so ora nella veglia continua del mio sonno,
ma l'ho capito nel momento di cui ho ora memoria
Umberto Eco, writer

Aharonov Y., Bohm D. (1959), Significance of electromagnetic potentials in the quantum theory, Phys. Rev., **115**, pp. 485–491.

Aharonov Y., Bohm D. (1961), Further considerations on electromagnetic potentials in the quantum theory, Phys. Rev., **123**, pp. 1511–1524.

Alanakyan Y. R. (1994), Energy capacity of an electromagnetic vortex in the atmosphere, JETP, **78**, p. 320.

Aldrovandi R., Pereira J. G., Vu K. H. (2007), The nonlinear essence of gravitational waves, Found. Phys., **37**, pp. 1503–1517.

Anderson E. (2004), Geometrodynamics, spacetime or space?, arXiv:gr-qc/0409123v1.

Arbab I. (2017), Extended electrodynamics and its consequences, Mod. Phys. Lett. B, **31**, 9, 1750099.

Arcos H. I., Pereira J. C. (2004), Kerr-Newman solution as a Dirac particle, Gen. Rel. Grav., **36**, 11, pp. 2441–2464.

Ardavan H. (1984), A singularity arising from the coherent generation of gravitational waves by electromagnetic waves, in *Classical General Relativity*, Bonner W. B., Islam I. N., MacCollum M. A. H. (eds.), Cambridge Univ. Press.

Ashkin A. (1969), Acceleration and trapping of particles by radiation pressure, Phys. Rev. Lett., **24**, 4, pp. 156–159.

Atkins P. W. (1990), *Physical Chemistry*, Oxford Univ. Press.

Atwater R. A. (1994), *Basic Relativity*, Springer.

Awada A., Alia A. F., Majumderd B. (2013), Nonsingular rainbow universes,

JCAP, October.

Bader R. F. W. (1991), A quantum theory of molecular structure and its application, Chem. Rev., **91**, pp. 893–928.

Bader R. F. W. (1994), *Atoms in Molecules, A Quantum Theory*, Clarendon Press.

Badiale M., Benci V., Rolando S. (2004), Solitary waves: physical aspects and mathematical results, Rend. Sem. Mat. Univ. Pol. Torino, **62**, 2, pp. 107–154.

Bahder B., Fazi C. (2002), Force on an asymmetric capacitor, arXiv:physics/0211001v2.

Bailar J. C., Eméléus H. J., Sir Nyholm R. S., Trotman-Dickenson A. F. (eds.) (1973), *Comprehensive Inorganic Chemistry*, Pergamom Press.

Bajpay R. P. (2015), Biophotonic route for understanding mind, brain and the world, Cosmos and History: The Journal of Natural and Social Philosophy, **11**, 2, pp. 189–199.

Bak P., Tang C., Wiesenfeld K. (1988), Self-organized criticality, Phys. Rev. A, **38**, 1, pp. 364–374.

Barabási A.-L., Stanley H. E. (1995), *Fractal Concepts in Surface Growth*, Cambridge Univ. Press.

Barenghi C. F. Donnelly R. J. (2009), Vortex rings in classical and quantum systems, Fluid Dyn. Res., **41**, 051401.

Barrett T. W. (2008), *Topological Foundations of Electromagnetism*, SCCP, Vol. 26, World Scientific.

Barrow G. M. (1996), *Physical Chemistry*, McGraw-Hill.

Batchelor G. K. (1967), *An Introduction to Fluid Dynamics*, Cambridge Univ. Press.

Bearden T. (2004), *Energy from the Vacuum, Concepts & Principles*, Cheniere Press.

Beil R. G. (1997), A classical photon model, in *Present Status of the Quantum Theory of Light*, Jeffers S., Roy S., Vigier J.-P., Hunter G. (eds.), Kluwer Academic pp. 9–16.

Benci V., Fortunato D. (2002), Solitary waves of the nonlinear Klein-Gordon field equation coupled with the Maxwell equations, Rev. Math. Phys., **14**, pp. 409–420.

Benci V., Fortunato V. (2004), Towards a unified field theory for classical electrodynamics, Arch. Rat. Mech. Anal., **173**, pp. 379–414.

Benci V., Fortunato D., Masiello A., Pisani L. (1999), Solitons and the electromagnetic field, Math. Z., **232**, pp. 73–102.

Benson T. M. et al. (2005), Micro-optical resonators for microlasers and integrated optoelectronics: Recent advances and future challenges, in *Frontiers of Planar Lightwave Circuit Technology, Simulation and Fabrication*, Janz S., Ctyroky J., Tanev S. (eds.), Springer, pp. 39–70.

Bergmann P. G. (1968), *The Riddle of Gravitation*, Charles Scribner's Sons.

Bertone G. (editor) (2014), *Particle Dark Matter: Observations, Models and Searches*, Cambridge Univ. Press.

Bigelow M. S., Zerom P., Boyd R. W. (2004), Breakup of ring beams carrying orbital angular momentum in sodium vapor, Phys. Rev. Lett., **92**, 8, pp. 083902.

Biró L. P. et al. (2007), Living photonic crystals: Butterfly scale – Nanostructure and optical properties, Mater. Sci. Eng. C, **27**, pp. 941–946.

Böhm A. (1979), *Quantum Mechanics*, Springer.

Bohm D. (1957), *Casuality and Chance in Modern Physics*, Van Nostrand.

Bohm D. J., Hiley B. J. (1982), The de Broglie pilot wave theory and the further development of new insights arising out of it, Found. Phys., **12**, 10, pp. 1001–1016.

Bokulich A. (2008), *Reexamining the Quantum-Classical Relation*, Cambridge Univ. Press.

Bondi H., Pirani F. A. E., Robinson I. (1959), Gravitational waves in general relativity, III, Exact plane waves, Proc. R. Soc. Lond. A, **144**, pp. 425–451.

Boriskina S. V. (2007), Coupling of whispering-gallery modes in size-mismatched microdisk photonic molecules, Optics Letters, **32**, 11, pp. 1557–1559.

Boriskina S. V. (2010), Photonic molecules and spectral engineering, in *Photonic Microresonator Research and Applications*, Chremmos I., Uzunoglu N., Schwelb O. (eds.), Springer, pp. 393–421.

Born M., Infeld L. (1934), Foundations of the new field theory, Proc. R. Soc. Lon. A, **144**, pp. 425–451.

Born M., Wolf E. (1980), *Principles of Optics*, 4th ed., Pergamon Press.

Bostick W. H. (1985), The morphology of the electron, Int. J. of Fusion Energy, **3**, 1, pp. 9–50.

Boyer T. H. (1968), Quantum electromagnetic zero-point energy of a conducting spherical shell and the Casimir model for a charged particle, Phys. Rev., **174**, 5, pp. 1764–1776.

Boyer T. H. (1975), Random electrodynamics: The theory of classical electrodynamics with classical electromagnetic zero-point radiation, Phys. Rev. D, **11**, 4, pp. 790–808.

Brady D. A. (2014), Anomalous thrust production from an RF test device measured on a low-thrust torsion pendulum, 50th AIAA/ASME/SAE/ASEE Joint Propulsion Conference, AIAA Propulsion and Energy Forum, 2014-4029.

Bryce M., Balick B., Meaburn J. (1994), Investigating the haloes of planetary nebulae, Part four – NGC6720 the ring nebula, R.A.S. Monthly Notices, **266**, p. 721.

Budden K. G., Martin H. G., Mott N. F. (1962), The ionosphere as a whispering gallery, Proc. Roy. Soc. Ser. A, **265**, pp. 554–569.

Burcham W. E., Jobes M. (1997), *Nuclear and Particle Physics*, Longman.

Bychkov V. L. et al. (eds.) (2014), *The atmosphere and Ionosphere, Elementary Processes, monitoring and Ball Lightning*, Springer.

Canning F. X., Melcher C., Winet E. (2004), Asymmetrical capacitors for propulsion, NASA report: CR-2004-213312.

Cantore E. (1969), *Atomic Order, An Introduction to the Philosophy of Microphysics*, The MIT Press.

Capella B, Dietler G. (1999), Force-distance curves by atomic force microscopy,

Surf. Sci. Rep., **34**, pp. 1–104.

Casimir H. B. G. (1948), On the attraction between two perfectly conducting plates, Proc. Kon. Nederland. Akad. Wetensch., **B51**, p. 793.

Casimir H. B. G., Polder D. (1948), The influence of retardation on the London–van der Waals forces, Phys. Rev., **73**, pp. 360–372.

Cattani F. et al. (2013), Interactions of electromagnetic radiation with Bose-Einstein condensates: Manipulating ultra-cold atoms with light, Int. J. Mod. Phys. B, **27**, 6, 1330003.

Chen C., Pakter R., Seward D. C. (2001), Equilibrium and stability properties of self-organized spiral toroids, Phys. Plasmas, **8**, 10, pp. 4441–4449.

Chinosi C., Della Croce L., Funaro D. (2010), Rotating electromagnetic waves in toroid-shaped regions, Int. J. Mod. Phys. C, **21**, 1, pp. 11–32.

Chupp T.E., Fierlinger P., Ramsey-Musolf M.J., Singh J.T. (2019), Electric dipole moments of atoms, molecules, nuclei, and particles, Rev. Mod. Phys., **91**, 015001.

Cifra M., Fields, J. Z., Farhadi A. (2011), Electromagnetic cellular interactions, Prog. Biophys. Mol. Biol., **105**, pp. 223–246.

Clauser F. H. (editor) (1960), *Symposium on Plasma Dynamics*, Addison-Wesley.

Clemmow P. C. (1966), *The Plane Wave Spectrum Representation of Electromagnetic Fields*, Pergamon Press.

Coclite G. M., Georgiev V. (2004), Solitary waves for Maxwell Schrödinger equations, Electronic J. Differ. Eq., **94**, pp. 1–31.

Cole D. C., Puthoff H. E. (1993), Extracting energy and heat from the vacuum, Phys. Rev. E, **48**, 2, pp. 1562–1565.

Cornille P. (2004), *Advanced Electromagnetism and Vacuum Physics*, SCCP, Vol. 21, World Scientific.

Cosic I. et al. (2006), Human electrophysiological signal responses to ELF Schumann resonance and artificial electromagnetic fields, FME Transactions, **34**, 2, pp. 93–103.

Cravens T. E. (1997), *Physics of Solar System Plasmas*, Cambridge Univ. Press.

Dawson G. A., Jones R. C. (1969), Ball lightning as a radiation bubble, Pure Appl. Geophys, **75**, pp. 247–262.

Dayton B. B. (2012), Hydrodynamic model of pions, muons, kaons, and the strong force, Phys. Essays, **25**, 1, pp. 8–26.

De Benedetti S. (1964), *Nuclear Interactions*, John Wiley & Sons.

de la Peña L., Cetto A. M. (1966), *The Quantum Dice, An Introduction to Stochastic Electrodynamics*, Kluwer Academic.

Desiraju G. R. (2011), A bond by any other name, Angew. Chem., Int. Ed., **50**, pp. 52–59.

Deutsch I. H., Chiao R. Y., Garrison J. C. (1992), Diphotons in a nonlinear Fabry–Pérot resonator: Bound states of interacting photons in an optical "quantum wire", Phys. Rev. Lett., **69**, 25, pp. 3627–3630.

Dewey E. R., Dakin E. F. (1947), *Cycles, the Science of Prediction*, H. Holt & Company.

Dirac P. A. M., Fowler R. H. (1928), The quantum theory of the electron, Proc. Roy. Soc. London, **A 117**, pp. 610–624.

Dirac P. A. M. (1967), *The Principles of Quantum Mechanics*, Oxford Univ. Press.

Dodonov V. V., Dodonov A. V. (2005), Quantum harmonic oscillator and non-stationary Casimir effect, J. Russ. Laser Res., **26**, 6, pp. 445–483.

Domon K., Ishihara O., Watanabe S. (2000), Mass transport by a vortex ring, J. Phys. Soc. Jpn., **69**, pp. 120–123.

Donev S., Tashkova M. (2012), Geometric view on photon-like objects, arXiv: 1210.8323v4.

Dragoman D., Dragoman M. (2004), *Quantum-Classical Analogies*, Springer.

Dyson F. J. (1972), Missed opportunities, Bull. Amer. Math. Soc., **78**, 5, pp. 635–652.

Efremidis N. K., Hizanidis K., Malomed B. A., Di Trapani P. (2007), Three dimensional vortex solitons in self-defocusing media, Phys. Rev. Lett., **98**, 113901.

Einstein A. (1905), On the electrodynamics of moving bodies, English translation from the original paper "Zur Elektrodynamik bewegter Körper", Annalen der Physik., **17**, pp. 891–921.

Einstein A. (1920), English translation from the original paper, 1916, *Relativity: The Special and General Theory*, H. Holt & Company.

Eisberg R., Resnick R. (1985), *Quantum Physics of Atoms, Molecules, Solids, Nuclei, and Particles*, John Wiley & Sons.

Elcrat A. R., Fornberg B., Miller K. G. (2001), Some steady axissymmetric vortex flows past a sphere, J. Fluid Mech., **433**, p. 315.

Elhakeem et al. (2018), Aboveground mechanical stimuli affect belowground plant-plant communication, PLoS One, **13**, 5, e0195646.

Elitzur A., Dolev S., Kolenda N. (eds.) (2005), *Quo Vadis Quantum Mechanics?*, Springer.

Endean V. G. (1976), BL as electromagnetic radiation, Nature, **263**, pp. 753–755.

Engelhardt W. (2012), On the Solvability of Maxwell's Equations, Ann. Fond. Louis de Broglie, **37**, pp. 3–14.

Engels S. et al. (2014), Anthropogenic electromagnetic noise disrupt magnetic compass orientation in migratory bird, Nature, **509**, p. 353.

Evans M., Vigier J. P. (1994), *The Enigmatic Photon*, Vol. 1 and 2, Kluwer Academic.

Farina C. (2006), The Casimir effect: some aspects, Braz. J. Phys., **36**, 4A, pp. 1137–1149.

Feather N. (1968), *Electricity and Matter*, Edinburgh Univ. Press.

Feynman R. P. (1985), *QED, The Strange Theory of Light and Matter*, Princeton Univ. Press.

Feynman R. P. (1998), *Quantum Electrodynamics*, Westview Press.

Feynman R. P., Leighton R. B., Sands M. (1963), *The Feynman Lectures on Physics*, Addison Wesley.

Filipponi G. (1985), The importance of Eugenio Beltrami's Hydroelectrodynamics, Int. J. Fusion Energy, **3**, 3, pp. 37–57.

Finkelnburg W. (1964), *Structure of Matter*, Springer.

Firstenberg O. et al. (2013), Attractive photons in a quantum nonlinear medium, Nature, **502**, pp. 71–75.

Firth W. J., Skryabin D. V. (1997), Optical solitons carrying orbital angular momentum, Phys. Rev. Lett., **79**, 13, pp. 2450–2453.

Fischer A. M. et al. (2009), Exciton storage in a nano-scale Aharonov-Bohm ring with electric field tuning, Phys. Rev. Lett., **102**, 9, 096405.

Fleming T. (2014), *Inside the Photon: A Journey to Health*, Pan Stanford.

Flick J., Ruggenthaler M., Appel H., Rubio A. (2017), Atoms and molecules in cavities, from weak to strong coupling in quantum-electrodynamics (QED) chemistry, PNAS, **114**, 12, pp. 3026–3034.

Fock V. (1959), *The Theory of Space, Time and Gravitation*, Pergamon Press, London.

Ford L. H. (1988), Spectrum of the Casimir effect, Phys. Rev. D, **38**, 2, pp. 528–532.

Ford L. H. (2007), Frequency spectra and probability distributions for quantum fluctuations, Int. J. Theor. Phys., **46**, pp. 2218–2226.

Fritzsch H. (1983), *Quarks, The Stuff of Matter*, Allen Lane.

Funaro D. (2005), A full review of the theory of electromagnetism, arXiv: physics/0505068.

Funaro D. (2008), *Electromagnetism and the Structure of Matter*, World Scientific.

Funaro D. (2009a), The fractal structure of matter and the Casimir effect, arXiv:0906.1874v1.

Funaro D. (2009b), Electromagnetic radiations as a fluid flow, arXiv:0911.4848v1.

Funaro D. (2010), Numerical simulation of electromagnetic solitons and their interaction with matter, J. Sci. Comput., **45**, pp. 259–271.

Funaro D. (2010), A Lagrangian for electromagnetic solitary waves in vacuum, arXiv:1008.2103v1.

Funaro D. (2014), Charging capacitors according to Maxwell's equations: impossible, Ann. Fond. Louis de Broglie, **39**.

Funaro D. (2014), Trapping electromagnetic solitons in cylinders, Math. Model. Anal., **19**, 1, pp. 44–51.

Funaro D. (2018a), A Model for ball lightning derived from an extension of the electrodynamics equations, Proc. VI Int. Conf. on Atmosphere, Ionosphere, Safety, Kaliningrad.

Funaro D. (2018b), High frequency electrical oscillations in cavities, Math. Model. Anal., **23**, 3, pp. 345–358.

Funaro D., Kashdan E. (2015), Simulation of electromagnetic scattering with stationary or accelerating targets, Int. J. Mod. Phys. C, **26**, 7, pp. 1–16.

Gadre S. R., Shirsat R. N. (2000), *Electrostatic of Atoms and Molecules*, Universities Press, Hyderabad.

Gallavotti G. (2002), *Foundations of Fluid Dynamics*, Springer.

Geim A. (1998), Everyone's magnetism, Physics Today, **51**, 9, pp. 36–39.

Gillespie R. J. (1972), *Molecular Geometry*, Van Nostrand.

Gladman B., Quinn D. D., Nicholson P., Rand R. (1996), Synchronous locking of tidally evolving satellites, Icarus, **122**, 1, pp. 166–192.

Gomes L. F. et al. (2013), Shapes and vorticities of superfluid helium nanodroplets, Science, **345**, 6199, pp. 906–909.

Gould S. J. (2002), *The Structure of Evolutionary Theory*, Harvard Univ. Press.

Gross D. (2005), Einstein and the search for unification, Current Science, **98**, 12, pp. 2035–2040.

Guinier A. (1984), *The Structure of Matter*, Edward Arnold.

Haken H. (1981), *Light*, Vol. 1, North–Holland.

Haken H., Wolf H. C. (1994), *The Physics of Atoms and Quanta*, Springer.

Harmuth H. F., Barrett T. W., Meffert B. (2001), *Modified Maxwell Equations in Quantum Electrodynamics*, SCCP, Vol. 19, World Scientific.

Hasimoto H. (1972), A soliton on a vortex filament, J. Fluid Mech., **51**, pp. 477–485.

Hawking S., Israel W. (eds.) (1987), *300 Years of Gravitation*, Cambridge Univ. Press.

Hazeltine R. D., Meiss J. D. (2003), *Plasma Confinement*, Dover.

He Y. J., Malomed B. A., Wang H. Z. (2007), Steering the motion of rotary solitons in radial lattices, Phys. Rev. A, **76**, 053601.

Hecht R. (2002), *Optics*, Addison-Wesley, International Edition.

Heitler W., London F. (1927), Wechselwirkung neutraler atome und homöpolare bindung nach der quantenmechanik, Zeitschrift für Physik, **44**, 6, 7, pp. 455–472.

Heitler W. (1944), *The Quantum Theory of Radiation*, 2nd ed., Oxfor Univ. Press.

Henderson G. (1980), Quantum dynamics and a semiclassical description of the photon, Am. J. of Phys., **48**, p. 604.

Hendry E. (2011), *The Metaphysics of Chemistry*, Oxford Univ. Press.

Hively L. M., Giakos G. C.. (2012), Toward a more complete electrodynamic theory, IJSISE, **5**, 1, pp. 3–10.

Hoddeson L., Brown L., Riordan M., Dresden M. (eds.) (1997), *The Rise of the Standard Model, A History of Particle Physics from 1964 to 1979*, Cambridge Univ. Press.

Horspool W. M., Song P.-S. H. (eds.) (1955), *CRC Hahdbook of Organic Photochemistry and Photobiology*, CRC Press.

Hsueh C.-H., Gou S.-C., Horng T.-L., Kao Y.-M. (2007), Vortex-ring solutions of the Gross–Pitaevskii equation for an axisymmetrically trapped Bose–Einstein condensate, J. Phys. B, **40**, pp. 4561–4571.

Humphries S. Jr. (1990), *Charged Particle Beams*, Wiley-Interscience.

Hunter G., Wadlinger R. L. P. (1989), Photons and neutrinos as electromagnetic solitons, Phys. Essays, **2**, pp. 158–172.

Hunter G. (1997), Electrons and photons as soliton waves, in *Present Status of the Quantum Theory of Light*, Jeffers S., Roy S., Vigier J.-P., Hunter G. (eds.), Kluwer Academic, pp. 37–44.

Ibison M., Haisch B. (1996), Quantum and classical statistics of the electromagnetic zero-point field, Phys. Rev. A, **54**, 4, pp. 2737–2744.

Ijjas A., Steinhardt P. J., Loeb A. (2017), POP goes the universe, Sci. Am., **32**.

Innis R. E. (1994), *Consciousness and the Play of Signs*, Indiana Univ. Press.

Irvine W. T. M., Bouwmeester D. (2008), Linked and knotted beams of light, Nature Phys., **4**, pp. 716–719.

Jackson J. D. (1975), *Classical Electrodynamics*, 2nd ed., John Wiley & Sons.

Jaffe R. L. (1995), Where does the proton really gets its spin?, Physics Today, Sept., pp. 24–30.

Kaku M. (1991), *String, Conformal Fields and Topology, An Introduction*, Springer-Verlag.

Kamor A. et al. (2012), Annular billiard dynamics in a circularly polarized strong laser field, Phys. Rev. E, **85**, 016204.

Karplus M., Porter R. N. (1970), *Atoms and Molecules, an Introduction for Students of Physical Chemistry*, W.A. Benjamin.

Karzig T. et al. (2015), Topological polaritons, Phys. Rev. X, **5**, 031001.

Kedia H. et al. (2013), Tying knots in light fields, Phys. Rev. Lett., **111**, 150404.

Keller O. (2005), On the theory of spatial localization of photons, Phys. Rep., **411**, pp. 1–232.

Kenneth O., Klich I., Mann A., Revzen M. (2002), Repulsive Casimir forces, Phys. Rev. Lett. **89**, 033001.

Kevrekidis P. G. et al. (2001), Ring solitons on vortices, Phys. Rev. E, **64**, 066611.

Kidd R., Ardini J., Anton A. (1989), Evolution of the modern photon, Am. J. Phys., **57** pp. 27–35.

Kittel C. (1962), *Introduction to Solid State Physics*, 2nd ed., John Wiley & Sons.

Kline M., Kay I W. (1965), *Electromagnetic Theory and Geometric Optics*, John Wiley & Sons.

Kramers H. A., Holst H. (1923), *The Atom and the Bohr Theory of its Structure, An Elementary Presentation*, London, Gyldendal.

Kruglov V. I., Vlasov R. A. (1985), Spiral self-trapping propagation of optical beams in media with cubic nonlinearity, Phys. Lett., **111A**, 8-9, pp. 401–404.

Kuo H.-C., Lin L.-Y., Chang C.-P., Williams R. T. (2004), The formation of concentric vorticity structures in typhoons, J. Atmos. Sci., **61**, pp. 2722–2734.

Landau L., Lifshitz E. (1961), *The Classical Theory of Fields*, Pergamon Press.

Leach A. R. (2001), *Molecular Modelling, Principles and Applications*, 2nd ed., Pearson Education.

Lee S., Lee Y., Yu I. (2005), Electric field in solenoids, Jpn. J. Appl. Phys., **44**, 7A, pp. 5244–5248.

Leedskalnin E. (1945), *Magnetic Current*, personal copyright, Homestead FL.

Lehnert B., Roy S. (1998), *Extended Electromagnetic Theory: Space Charge in Vacuo and the Rest Mass of the Photon*, SCCP, Vol. 16, World Scientific.

Leibniz G. W. (1714), *The Monadology*, English translation by R. Latta, 1898, Oxford Univ. Press.

Liang Q.-Y. et al. (2018), Observation of three-photon bound states in a quantum nonlinear medium, Science, **359**, 6377, pp. 783–786.

Lim T. T., Nickels T. B. (1992), Instability and reconnection in the head-on collision of two vortex rings, Nature, **357**, pp. 225–227.

Lim T. T., Nickels T. B. (1995), Vortex rings, in *Fluid Vortices*, Green S. I. (editor), Kluwer Academic.

Linden P. F., Turner J. S. (2001), The formation of 'optimal' vortex rings, and the efficiency of propulsion devices, J. Fluid Mech., **427**, p. 61.

Linnett J. W. (1945), *The electronic Structure of Molecules, A New Approach*, Methuen (London), J. Wiley & Sons (New York).

Lipkowski J., Ross P. N. (eds.) (1999), *Imaging of Surface and Interfaces*, Wiley-VCH.

Little L. E. (1996), The theory of elementary waves, Phys. Essays, **9**, pp. 100–132.

Little L. E. (2009), *The Theory of Elementary Waves: A New Explanation of Fundamental Physics*, New Classics Library.

Lo C. Y. (2006), The gravity of photons and the necessary rectification of Einstein equation, Prog. Phys., **1**, pp. 46–51.

Lorenz W. J., Plieth W. (eds.) (1998), *Electrochemical Nanotechnology: In-situ Local probe Techniques at Electrochemical Interfaces*, Wiley-VCH.

Lothian G. F. (1963), *Electrons in Atoms*, Butterworths.

Loudon R. (2000), *The Quantum Theory of Light*, Oxford Univ. Press.

Low F. E. (1975), Model of the bare pomeron, Phys. Rev. D, **12**, 1, pp. 163–173.

Ma D., Stoica A. D., Wang X.-L. (2009), Power–law scaling and fractal nature of medium-range order in metallic glasses, Nature Materials, **8**, pp. 30–34.

Mabuchi H., Kimble H. J. (1993), Atom galleries for whispering atoms: Binding atoms in stable orbits around an optical resonator, Optics Lett., **19**, 10, pp. 749–751.

Mac Gregor M. H. (1992), *The Enigmatic Electron*, Kluwer Academic, Dordrecht.

Mandelbrot B. (1982), *The fractal Geometry of Nature*, Freeman.

Marani M. et al. (1998), Stationary self-organized fractal structures in an open, dissipative electrical system, J. Phys. A: Math. Gen., **31**, L337.

Marshak R. E. (1993), *Conceptual Foundations of Modern Particle Physiscs*, World Scientific.

Martin B. R., Shaw G. (2008), *Particle Physics*, 3rd ed., J. Wiley & Sons.

Maruani J. (1988), *Molecules in Physics, Chemistry and Biology*, Kluwer Academic.

Mathews C. K., van Holde K. E. (1990), *Biochemistry*, The Benjamin Cummings Publishing Company.

Maxwell J. C. (1861), On physical lines of forces, The London, Edinb. Dubl. Phil. Mag., pp. 161–175, 281–291.

Maxwell J. C. (1865), A dynamical theory of electromagnetic fields, Philos. Trans. R. Soc. Lond., **155**, pp. 459–512.

Maxwell Commemorative Booklet (1999), produced by the J. C. Maxwell Foundation on the occasion of the Fourth Int. Congress on Industrial and Applied Math., Edinburgh.

Maxworthy T. J. (1972), The structure and stability of vortex rings, J. Fluid Mech., **51**, 1, pp. 15–32.

McDonald K. T. (2014), Radiation in the near zone of a center-fed linear antenna, Preprint, Princeton University, updated in 2018.

Meis C. (2017), *Light and Vacuum*, 2nd ed., World Scientific.

Meis C., Dahoo P. R. (2017), Vector potential quantization and the photon intrinsic electromagnetic properties, Int. J. of Quantum Information, **15**, 8, 1740003.

Mezey P. G. (1993), *Shape in Chemistry, An Introduction to Molecular Shape and Topology*, VCH Pub., New York.

Miller A. I. (1984), *Imagery in Scientific Thought*, Birkhäuser.

Miller G. A. (2008), Non-spherical shapes of the proton: existence, measurement and computation, Nucl. Phys. News, **18**, pp.12–16.

Milonni P. W. (1993), *The Quantum Vacuum: An Introduction to Quantum Electrodynamics*, Academic Press.

Milton K. A. (2001), *The Casimir Effect, Physical Manifestations of Zero-Point Energy*, World Scientific.

Minini F. (1985), *L'etere, Eppur si Muove*, Tipografia Grosso, Bra, Italy.

Misner C. W., Thorne K. S., Wheeler J. A. (1973), *Gravitation*, W. H. Freeman & c..

Modanese G. (2013), A comparison between the YBCO discharge experiments by E. Podkletnov and C. Poher, and their theoretical interpretations, arXiv:1312.0958.

Moyroud E. et al. (2017), Disorder in convergent floral nanostructures enhances signalling to bees, Nature, 24285.

Munday J. N., Capasso F. (2007), Precision measurements of the Casimir–Lifshitz force in a fluid, Phys. Rev. A, **75**, 060102(R).

Munz C.-D., Schneider R., Sonnendrücker E., Voss U. (1999), Maxwell's equations when the charge conservation is not satisfied, C. R. Acad. Sci. Paris, **328**, Série I, pp. 431–436.

Nambu Y. (1985), *Quarks, Frontiers in Elementary Particle Physics*, World Scientific.

Nernst W. (1916), Über einen Versuch, von quantentheoretischen Betrachtungen zur Annahme stetiger Energieänderungen zurückzukehren, Verhandlungen der Deutschen Physikalischen Gesellschaft, **18**, 4 pp. 83–116.

Newton R. G. (2009), *How Physics Confronts Reality*, World Scientific.

Nikitin A. I. (2004), The principles developing ball lightning theory, J. Russian Laser Res., **25**, pp. 169–191.

Nikolić H. (2007), Quantum mechanics: Myths and facts, Found. Phys., **37**, pp. 1536–1611.

Nussinov S. (1975), Colored-quark version of some hadronic puzzles, Phys. Rev. Lett., **34**, 20, pp. 1286–1289.

Okun L. B. (1987), *Leptons and Quarks*, North Holland.

Oppenheimer R. J. (1989), *Atom and Void*, Princeton Univ. Press.

Oraevsky A. N. (2002), Whispering-gallery waves, Quantum Electronics, **32**, pp. 377–400.

Panarella E. (1986), Effective photon hypothesis vs. quantum potential theory: Theoretical predictions and experimental verification, in *Quantum Uncertainties*, Honig W. M., Kraft D. W., Panarella E. (eds.), NSSB, **162**, Plenum Press, pp 237–269.

Papasimakis N. et al. (2016), Electromagnetic toroidal excitations in matter and free space, Nature Materials, **15**, pp. 263–271.

Pauling L. (1960), *The Nature of the Chemical Bond*, Cornell Univ. Press.

Peitgen H.-O., Jürgens H., Saupe D. (1992), *Chaos and Fractals, New Frontiers of Science*, Springer.

Pellegrini G. N., Swift A. R. (1995), Maxwell's equations in a rotating medium: Is there a problem?, Am. J. Phys., **63**, pp. 694–705.

Piazza L. et al. (2015), Simultaneous observation of the quantization and the interference pattern of a plasmonic near-field, Nature Communications, **6**, 6407.

Planck M. (1927), The physical reality of light-quanta, Jour. Frank. Inst., pp. 13–18.

Podkletnov E., Modanese G. (2003), Investigation of high voltage discharges in low pressure gases through large ceramic superconducting electrodes, J. Low Temp. Phys., **132**, pp. 239–259.

Poher C., Poher D. (2011), Physical phenomena observed during strong electric discharges into layered Y123 superconducting devices at 77 K, Appl. Phys. Res., **3**, pp. 51–66.

Poincaré M. H. (1960), Sur la dynamique de l'électron, Rend. Circ. Palermo, **XXI**, pp. 129–175.

Popp F.-A. (2003), Properties of biophotons and their theoretical implications, Indian J. Exp. Biol., **41**, pp. 391–402.

Popper K. R. (1982), *Quantum Theory and the Schism in Physics*, Hutchinson.

Preston M. A. (1962), *Physics of the Nucleus*, Addison-Wesley.

Pullin D. I. (1979), Vortex ring formation at tube and orifice openings, Phys. Fluids, **22**, p. 401.

Pullman A., Dreyfus M., Mély B. (1970), Aspects of the electron distribution in adenine, thymine and cytosine as given by probability density curves from nonempirical calculations, Theoret. Chim. Acta (Berl.), **16**, pp. 85–11.

Rañada A. F., Trueba J. L. (1996), Ball lightning an electromagnetic knot?, Nature, **383**, pp. 32.

Raychaudhuri P. (1986), Compositeness of photons and its implications, in *Quantum Uncertainties*, Honig W. M., Kraft D. W., Panarella E. (eds.), NSSB, **162**, Plenum Press, pp. 271–284.

Ricci G., Ruggiero M. L. (2002), Space geometry of rotating platforms: An operational approach, Found. Phys., **32**, 10, pp. 1525–1556.

Ricci G., Ruggiero M. L. (2004), *Relativity in Rotating Frames*, Fundamental Theories of Physics, Vol. 135, Springer.

Rice M.H., Good R.H. Jr.(1962), Stark effect in hydrogen, J. Opt. Soc. Amer., **52**, 3, pp. 239–246.

Rice F. O., Teller E. (1949), *The Structure of Matter*, John Wiley & Sons.

Roychoudhuri C., Kracklauer A. F., Creath K. (eds.) (2008), *The Nature of Light: What is a Photon?*, CRC Press.

Sachs M. (2004), *Quantum Mechanics and Gravity*, Springer.

See T. F. F. (1920), New theory of aether, Astron. Nachr., Band 211, 5044, pp. 49–86.

Segrè E. (1977), *Nuclei and Particles, An Introduction to Nuclear and Subnuclear Physics*, 2nd ed., W. A. Benjamin.

Sezginer A. (1985), A general formulation of focus wave modes, J. Appl. Phys., **57**, pp. 678–683.

Shariff K., Leonard A. (1992), Vortex rings, Annual Rev. Fluid Mech., **24**, p. 235.

Sharpe A. G. (1981), *Inorganic Chemistry*, Longman.

Shen J.-T., Fan S. (2007), Strongly correlated two-photon transport in a one-dimensional waveguide coupled to a two-level system, Phys. Rev. Lett., **98**, 153003.

Shikhmurzaev Y. D. (2008), *Capillar Flows with Forming Interfaces*, Chapman & Hall, Boca Raton FL.

Silliman R. H. (1963), William Thomson: Smoke rings and the nineteenth-century atomism, ISIS, **54**, 178, pp. 461–474.

Simulik V. (editor) (2005), *What is the Electron?*, Apeiron.

Snyder A., Love J. (1983), *Optical Waveguide Theory*, Kluwer Academic.

Stenhoff M. (2002), *Ball Lightning: An Unsolved Problem in Atmospheric Physics*, Kluwer Academic.

Stone J. M. (1963), *Radiation and Optics*, McGraw-Hill.

Sullivan I. S. et al. (2008), Dynamics of thin vortex rings, J. Fluid Mech., **609**, p. 319.

Tenforde T. S. (editor) (1979), *Magnetic Field Effect on Biological Systems*, Plenum Press.

Tikhonenko V., Christou J., Luther-Davies B. (1996), Three dimensional bright spatial soliton collision and fusion in a saturable nonlinear medium, Phys. Rev. Lett., **76**, 15, pp. 2698–2701.

Thomson J. J. (1935), The nature of light, Nature, Letters to the Editor, Feb. 8, pp. 232–233.

Thomson W. (Lord Kelvin) (1867), On vortex atoms, Proc. Royal Soc. Edinb., **VI**, pp. 94–105.

Tolhoek H. A. (1956), Electron polarization, theory and experiment, Rev. Mod. Phys., **28**, 3, pp. 277–298.

van der Merwel A., Garuccio A. (eds.) (1994), *Waves and Particles in Light and Matter*, Springer.

Van Dyke M. (1982), *An Album of Fluid Motion*, 10th edition, Parabolic Press.

Verlinde E. (2011), On the origin of gravity and the laws of Newton, JHEP, **29**.

Wakelin S. L., Riley N. (1997), On the formation of vortex rings and pairs of vortex rings, J. Fluid Mech., **332**, p. 121.

Weissbluth M. (2002), *Photon-Atom Interactions*, Elsevier.

Wheeler J. A. (1955), Geons, Phys. Rev., **97**, 2, pp. 511–536.

Wheeler J. A., Zurek W. H (eds.) (1983), *Quantum Theory and Measurement*, Princeton Univ. Press.

Williams W. S. C. (1991), *Nuclear and Particle Physics*, Clarendon Press.

Williamson J. G., van der Mark M. B. (1997), Is the electron a photon with toroidal topology?, Ann. Fond. Louis de Broglie, **22**, 2, pp. 133–158.

Woit P. (2006), *Not Even Wrong: The Failure of String Theory and the Search for Unity in Physical Law*, Basic Books.

Wooding E. R. (1963), Ball lightning, Nature, **199**, 4890, pp. 272–273.

Woodside R. W. M. (2004), Space-time curvature of classical electromagnetism, arXiv:gr-qc/0410043v1.

Xiao Y.-F. et al. (2010), High-Q exterior whispering-gallery modes in a metal-coated microresonator, Phys. Rev. Lett., **105**, 153902.

Yang Y. (2000), Classical solutions in the Born-Infeld theory, Proc. R. Soc. Lon. A, **456**, pp. 615–640.

Yang Y. (2000), *Solitons in Field Theory and Nonlinear Analysis*, Springer.

Yosida K. (1998), *Theory of Magnetism*, 2nd ed., Springer.

Zewail A. H. (2010), Filming the invisible in 4D: new microscopy makes movies of nanoscale objects in action, Sci. Am., **303**, 74.

Zhakatayev A. (2016), Divergence of electric field of continuous and of a point charge for relativistic and non-relativistic motion, arXiv:1608.02898.

Zhang T. L. et al. (2012), Magnetic reconnection in the near venusian magnetotail, Science, 1217013.

Ziman J. M. (1962), *Electrons and Phonons*, Oxford Univ. Press.

Zou You-Suo (1995), Some physical considerations for unusual atmospheric lights observed in Norway, Physica Scripta, **52**, p. 726.

Index